Practical Approach to Conducting a Systematic Review and Meta-Analysis

Abdelrahman M. Makram
https://orcid.org/0000-0003-2011-8092
Hessenklinik Stadtkrankenhaus Korbach, Germany

Nguyen Tien Huy
https://orcid.org/0000-0002-9543-9440
Nagasaki University, Japan

IGI Global
Scientific Publishing
Publishing Tomorrow's Research Today

Vice President of Editorial	Melissa Wagner
Director of Acquisitions	Mikaela Felty
Director of Book Development	Jocelynn Hessler
Production Manager	Mike Brehm
Cover Design	Jose Rosado

Published in the United States of America by
IGI Global Scientific Publishing
701 East Chocolate Avenue
Hershey, PA, 17033, USA
Tel: 717-533-8845 | Fax: 717-533-7115
Website: https://www.igi-global.com E-mail: cust@igi-global.com

Library of Congress Cataloging-in-Publication Data

No LCCN (CIP Data Pending)
ISBN13: 979-8-3693-3007-4
Isbn13Softcover: 979-8-3693-5190-1
EISBN13: 979-8-3693-3008-1

British Cataloguing in Publication Data
A Cataloguing in Publication record for this book is available from the British Library.

Table of Contents

Detailed Table of Contents

The primary purpose of this chapter is to present a helping guide for young and early-career researchers to navigate narrative reviews, knowing their concept, structure, advantages as well as their limitations. In addition, it presents a comprehensive practical step by step guide in writing and constructing an appropriate narrative synthesis-based review. Additionally, avoiding common pitfalls such as publication or researcher bias. The chapter mentions various real-world scenarios and case studies where the use of narrative synthesis helped in understanding and solving complex research questions, away from research or publication bias. In a style that allows research-based and other professionals/people to relate to. Another goal of this chapter is enhancing critical thinking for researchers by illustrating real-life examples in different settings such as concepts in randomised controlled trials.

This book chapter discusses the methods and applications of scoping and mapping reviews in medical research, explaining their distinct roles and how they relate to each other. Scoping and mapping reviews are both important exploratory tools designed to address broad research questions by summarizing a wide range of evidence. Scoping reviews are useful for providing an initial overview of the available literature, helping to clarify research questions and identify gaps in knowledge, especially for complex topics. In contrast, mapping reviews are effective for visually summarizing and categorizing existing literature, which helps identify themes and gaps in research. The chapter outlines the step-by-step processes necessary for conducting these reviews, from defining the research question to extracting data and presenting results. It also covers the advantages and challenges associated with each type of review, highlights the importance of involving stakeholders, and discusses the strategic use of different data extraction tools to improve the accuracy and relevance of the results.

Systematic reviews of diagnostic studies are essential tools for synthesizing evidence on the performance of diagnostic tests, providing clinicians and policymakers with valuable insights into test accuracy and reliability. Focusing on the accuracy of the diagnostic tests, these systematic reviews are commonly called the systematic reviews of diagnostic test accuracy (DTA). This chapter provides a detailed overview of these reviews, highlighting their specific methodological characteristics, types, and requirements and providing a thorough practical guide to conducting these studies. The key considerations for conducting a systematic review of diagnostic studies, including the formulation of the study question, protocol development, search strategy development, systematic search, study selection, data extraction, bias in DTA studies, critical appraisal, tools for risk of bias assessment, synthesis and meta-analysis, clinical utility, and implications for patient-oriented research are covered in this chapter.

Among different research methodologies, the umbrella review offers a very comprehensive tool for synthesizing evidence from multiple systematic reviews with or without meta-analyses. The umbrella review can provide a wider perspective by consolidating findings from various secondary evidence sources and highlighting the current state of evidence within a particular field. This chapter aims to discuss the detailed methodology of the umbrella review, its research applications, benefits, and challenges.

Scope and research questions must be the guiding light for the definition of the research methodology and the development of a review project. A well-defined scope ensures that research objectives remain focused and attainable by considering existing knowledge, its gaps, and potential novel contributions to the field. A rigorous formulation of a structured research question should follow a successful scope definition aiming at a meaningful contribution to the chosen field. While maintain the specificity of the research focus it's important to allow adjustments based on unforeseen challenges and emerging trends in research and policymaking, integrating them into the scope and research question while maintaining relevance and impact. To achieve this, stakeholder engagement should be integrated in research project planning. In summary, a meticulously crafted scope and research question are pillars of a review, fundamental to provide reliable information to research users applicable in a real setting to inform decision-making.

The initial phase for conducting SR/MA involves systematically searching multiple databases and locating all potentially relevant studies that could contribute to evaluating the review question. The reviewers establish parameters to skillfully control the search process and to specify which studies fit the systematic review's scope and which do not. These parameters are broadly known as "the eligibility criteria." Eligibility criteria can be further divided into inclusion criteria (to rule in studies) and exclusion criteria (to rule out studies). This fundamental step secures a high-quality research study, serving as the guiding principle for investigators to rely on and address their research questions effectively. This chapter reveals the role of inclusion and exclusion criteria in an SR/MA study. Besides discussing techniques for developing a good set of criteria, we will explore the common pitfalls to avoid, ensuring that the choices made during the design phase resonate with the broader context of research credibility.

Tailored for junior and beginner researchers in the medical research field, the chapter delves into the crucial process of screening articles for inclusion in systematic reviews and meta-analyses. This chapter serves as a comprehensive guide, addressing the fundamental aspects of selecting articles for inclusion in systematic reviews and meta-analyses. From exploring methods for establishing clear inclusion criteria to navigating challenges and outlining best practices, it equips researchers with the tools needed for a meticulous and reliable screening process to achieve impactful systematic reviews and meta-analyses.

This chapter serves as a comprehensive guide for researchers, especially junior and beginner researchers in the medical research field. Meta-analysis is explained in detail, emphasizing its role in synthesizing findings from various studies. Basic concepts such as effect sizes, effect models, and heterogeneity assessment are introduced, followed by an overview of the steps involved in conducting and interpreting a meta-analysis. Additionally, the chapter introduces advanced topics including subgroup analysis, meta-regression, and handling complex data structures. Practical recommendations and toolkits from the literature are provided to facilitate efficient meta-analysis. Furthermore, the chapter addresses current limitations and criticisms of meta-analyses, including issues of publication bias, heterogeneity, and subjectivity. By elucidating these challenges and offering potential solutions, the chapter aims to equip researchers with the necessary knowledge to conduct robust meta-analyses and contribute effectively to the field of medical research.

This chapter guides understanding and interpreting heterogeneity [within- and between-study variation] in meta-analyses, crucial for applying findings. We explain why heterogeneity matters and detail four key metrics: Cochran's Q, I^2, τ^2, and H statistic. For each, the chapter will clarify meaning, strengths, limitations,

Preface

INTRODUCTION

Systematic reviews and meta-analyses have become indispensable tools for synthesizing evidence and informing decision-making in medical research. By integrating findings from multiple studies, they offer a powerful means of addressing variability across the literature. Yet, despite their widespread use, the methodological foundations of these study designs remain the subject of ongoing debate. Divergent perspectives, concerns about information bias, and claims regarding their impracticality have led some to question their value altogether. We, however, firmly believe that when conducted rigorously and thoughtfully, systematic reviews and meta-analyses remain among the most effective approaches for managing data heterogeneity and generating meaningful, high-quality evidence.

While numerous resources describe the theoretical underpinnings of systematic reviews and meta-analyses, relatively few focus on the *practical* realities of conducting them. Researchers—particularly those at early stages of their careers—often find themselves overwhelmed by abstract methodological discussions, with limited guidance on how to translate theory into practice. This book was conceived to address that gap. Rather than reiterating what systematic reviews and meta-analyses are, we concentrate on *how* to conduct them. We emphasize practical steps, real-world considerations, common pitfalls, and actionable solutions. Throughout the chapters, readers will find tips, tricks, and recommended tools—many web-based—that can streamline workflows, improve efficiency, and support the production of robust, transparent, and reproducible reviews while minimizing the risk of information bias.

As editors, our goal is to provide a hands-on companion for researchers embarking on systematic review and meta-analysis projects. This book is intended for a broad audience within the medical research community, with particular attention to young researchers and undergraduate students. By breaking down complex processes into clear, manageable steps, we aim to demystify systematic reviews and meta-analyses, reduce apprehension, and build confidence. We hope this book will serve not only as a guide but also as an encouragement—helping readers navigate the process smoothly, anticipate challenges, and ultimately produce high-quality evidence that contributes meaningfully to medical science.

ORGANIZATION OF THE BOOK

Chapter 1: Introduction to Narrative Reviews: Narrative Synthesis

Recognizing that not all research questions lend themselves to quantitative synthesis, this chapter focuses on narrative reviews and narrative synthesis as valuable and often underappreciated approaches. Written with young and early-career researchers in mind, it introduces the concept, structure, advantages, and limitations of narrative reviews while offering a practical, step-by-step guide to conducting them rigorously. Through real-world scenarios and case studies, the chapter highlights how narrative synthesis can illuminate complex questions while minimizing bias and strengthening critical thinking across diverse research contexts.

Chapter 2: Exploring Scoping and Mapping Reviews

In this chapter, we explore scoping and mapping reviews as complementary exploratory approaches designed to address broad and complex research questions. The authors clearly distinguish between these methodologies, explaining when and why each should be used. A structured, stepwise guide walks readers through question formulation, data extraction, and result presentation, while also addressing methodological challenges and stakeholder engagement. By emphasizing visualization, categorization, and gap identification, this chapter demonstrates how scoping and mapping reviews can strategically inform future research agendas and policy discussions.

Chapter 3: Systematic Reviews of Diagnostic Studies

This chapter addresses the specialized yet critically important area of systematic reviews of diagnostic test accuracy (DTA). We provide a detailed and practical guide covering the unique methodological considerations required for these reviews, from formulating focused questions and developing protocols to assessing bias and synthesizing evidence. By discussing tools for critical appraisal, meta-analysis, and clinical interpretation, the chapter underscores how well-conducted DTA reviews directly inform patient-centered care and diagnostic decision-making.

Chapter 4: Beyond the Individual Review: A Practical Guide to Umbrella Review

As evidence synthesis continues to expand, umbrella reviews have emerged as powerful tools for summarizing findings across multiple systematic reviews. This chapter introduces the umbrella review as a method for providing a high-level, comprehensive perspective on entire fields of research. We discuss its methodology, applications, strengths, and limitations, offering readers practical guidance on how to navigate and integrate large bodies of secondary evidence while maintaining methodological rigor and transparency.

Chapter 5: Identifying Scope and Research Question

This chapter emphasizes that a well-defined scope and a carefully formulated research question are the cornerstones of any successful review. We discuss how thoughtful scoping aligns research objectives with existing evidence, gaps, and potential contributions, while structured research questions guide methodological choices. By highlighting flexibility, stakeholder engagement, and responsiveness to emerging trends, the chapter reinforces how these foundational decisions ensure relevance, rigor, and real-world applicability of review findings.

Chapter 6: Developing the Inclusion and Exclusion Criteria

Here, we focus on eligibility criteria as a critical design element that shapes the quality and credibility of systematic reviews and meta-analyses. The chapter explains how inclusion and exclusion criteria function as practical tools to control the search process and define the review's boundaries. Through practical guidance and discussion of common pitfalls, we demonstrate how well-constructed criteria strengthen methodological transparency and ensure that study selection aligns with the review's objectives.

Chapter 7: Screening Articles for Inclusion in SR/MA

Designed especially for junior and beginner researchers, this chapter provides a clear and practical roadmap for the screening process. We walk readers through best practices for applying eligibility criteria, managing challenges, and maintaining consistency and reliability during screening. By emphasizing meticulous methods and transparent decision-making, the chapter supports the production of systematic reviews and meta-analyses that are both rigorous and reproducible.

Chapter 8: How a Meta-Analysis Works

This chapter offers a comprehensive yet accessible introduction to meta-analysis, guiding readers from foundational concepts to more advanced techniques. Key elements such as effect sizes, models, and heterogeneity are explained alongside practical steps for conducting and interpreting analyses. We also address advanced topics, current criticisms, and practical solutions, ensuring readers are equipped not only to perform meta-analyses, but to critically appraise and responsibly interpret their results.

Chapter 9: Heterogeneity and Prediction Intervals in Meta-Analysis

The final chapter delves deeply into heterogeneity, one of the most challenging aspects of meta-analysis. We guide readers through key metrics—Cochran's Q, I^2, τ^2, and H—highlighting their complementary roles and limitations. Prediction intervals are introduced as a bridge between statistical findings and clinical relevance. By offering practical strategies for addressing heterogeneity and emphasizing transparent reporting, this chapter encourages a nuanced, responsible interpretation of meta-analytic evidence that goes beyond simplistic thresholds and supports informed decision-making.

CONCLUSION

As editors, we have brought this volume together with a clear and practical purpose: to guide researchers through the real-world process of conducting high-quality systematic reviews and meta-analyses with confidence, rigor, and transparency. Across the chapters, this book moves deliberately from foundational concepts to advanced methodological considerations, reflecting the natural progression of a well-designed evidence synthesis project. Each contribution was selected and shaped to bridge the persistent gap between methodological theory and day-to-day research practice.

This book does not present systematic reviews and meta-analyses as rigid or purely mechanical exercises. Rather, it emphasizes thoughtful decision-making, critical appraisal, and contextual judgment at every stage—from defining scope and questions, through study selection and synthesis, to interpretation and application of findings. By addressing diverse review types, statistical models, heterogeneity, and emerging approaches such as umbrella reviews, we aim to equip readers with both breadth and depth, enabling them to choose the most appropriate tools for their research questions.

We hope this volume serves not only as a reference, but also as a mentor-like companion for researchers at different stages of their careers.

For students and early-career investigators, it is intended to demystify systematic reviews and meta-analyses and reduce the anxiety often associated with conducting them. For experienced researchers, clinicians, and policymakers, it offers practical insights, methodological clarity, and reminders of best practices that strengthen the credibility and impact of evidence synthesis.

Ultimately, high-quality evidence synthesis is essential for advancing science, informing policy, and improving patient care. By emphasizing practical guidance, transparency, and critical thinking, we aspire for this book to contribute to a culture of rigorous, responsible, and meaningful research. We trust that readers will not only apply the methods presented here, but also carry forward the principles of methodological integrity and evidence-based decision-making that underpin this work.

Chapter 1
Introduction to Narrative Reviews:
Narrative Synthesis

Mohamed H. Dawood
https://orcid.org/0000-0002-9426-8256
Miami University, Oxford, USA

Queen Anoma
https://orcid.org/0009-0007-5312-5893
Imperial College London, UK

ABSTRACT

The primary purpose of this chapter is to present a helping guide for young and early-career researchers to navigate narrative reviews, knowing their concept, structure, advantages as well as their limitations. In addition, it presents a comprehensive practical step by step guide in writing and constructing an appropriate narrative synthesis-based review. Additionally, avoiding common pitfalls such as publication or researcher bias. The chapter mentions various real-world scenarios and case studies where the use of narrative synthesis helped in understanding and solving complex research questions, away from research or publication bias. In a style that allows research-based and other professionals/people to relate to.

DOI: 10.4018/979-8-3693-3007-4.ch001

Another goal of this chapter is enhancing critical thinking for researchers by illustrating real-life examples in different settings such as concepts in randomised controlled trials.

PURPOSE AND OBJECTIVES OF THIS CHAPTER

- This chapter aims to help young researchers by providing knowledge on narrative synthesis, the advantages, and disadvantages as well as the different approaches to take when conducting a narrative synthesis.
- Understanding the Principles of Narrative Synthesis: Students will gain a clear understanding of the underlying principles and methodologies of narrative synthesis, as well as its defining characteristics and research applications.
- Developing Skills in Data Integration and Interpretation: Through practical examples, students will learn how to systematically integrate and interpret diverse sources of evidence to construct cohesive research narratives.
- Enhancing Critical Thinking and Analytical Skills: This chapter will help students develop critical thinking skills to evaluate and synthesise evidence critically, identifying patterns, discrepancies, and implications for further investigation.
- Students will learn how to write clear, coherent, and persuasive narratives that convey synthesised findings and insights to a wide range of audiences, including academic colleagues, policymakers, and practitioners.

1. NARRATIVE SYNTHESIS: STORYTELLING TO ANSWER RESEARCH QUESTIONS

Qualitative research can take multiple forms, some of which are descriptive, and others are statistical and numerically based. There are multiple settings, where sophisticated meta-analysis approaches or thorough qualitative studies are not feasible, and that is when 'narrative syntheses can come in handy. Narrative synthesis is one of the primary techniques used in systematic reviews, characterised by relying on texts and words to elaborate or summarise the outcomes of a specific hypothesis/synthesis. The essence of narrative synthesis revolves around its exploration of various research questions concerning specific topic from different perspectives, moreover, narrative synthesis is distinguished from traditional narrative reviews by trying to combine various previous studies and their outcomes for a specific research question (Noyes et al., 2011; Sattar et al., 2021)

The core of the narrative approach is to enable researchers to uncover complexities and their associated findings, while also probing into the underlying causations behind the research question at hand by asking questions like "why" and "what." Thereupon, it can also be considered as one of the most inclusive forms of literature reviews; it builds up the knowledge from the considerably basic till the advanced tip which is usually the main topic theme "research question" (Erwin et al., 2011).

The core design of narrative synthesis revolves around presenting facts in a clear, logical, and bias-free structure. Besides summarising previous studies, narrative synthesis must also have a sense of critique in its body i.e. to interrogate the validity and quality of results from different studies (Doyle, 2003). In healthcare settings, narrative synthesis not only facilitates and answers a certain research question and its potential factors, but it also focuses on allowing researchers to, delve more into a particular topic and to gain insights about the positives and shortcomings of past methodologies in dealing with a particular topic (Erwin et al., 2011; Hannes & Macaitis, 2012).

Narrative synthesis, as introduced, is a methodology embedded within systematic reviews, offering a qualitative lens to amalgamate findings from multiple studies. This approach proves invaluable when quantitative meta-analysis encounters obstacles like heterogeneity or data scarcity. (Rodgers et al., 2009). Narrative synthesis combines statistical data manipulation with a textual approach to narrate study findings. It enables researchers to investigate complicated phenomena, comprehend contextual elements, and create rich insights that can inform practice, policy, and future research (Popay et al., 2006). At its foundation, narrative synthesis requires students to critically analyse the literature, identify patterns, connections, and discrepancies, and construct a cohesive story that synthesises multiple views and findings.

1.1 Historical Background and Evolution of Narrative Synthesis

Narrative synthesis, a component of qualitative research, falls within the broader scope of literature development. Originating in the early 1990s, it emerged as a methodological approach to systematically analyse and merge findings from diverse qualitative studies. Over time, narrative synthesis has undergone significant evolution. Its roots trace back to ancient philosophical traditions where researchers integrated various viewpoints. In the twentieth century, especially with the rise of systematic reviews in medical research, narrative synthesis emerged as a formal alternative to statistical methods like meta-analysis. This evolution extended synthesis techniques, notably incorporating qualitative evidence (Erwin et al., 2011; Laver et al., 2020).

Narrative synthesis gained prominence for its ability to integrate qualitative and quantitative data, focusing on literary interpretation rather than numerical aggregation. Methodological advancements introduced new strategies such as topic synthesis and realist synthesis. Widely used across disciplines, narrative synthesis remains valuable for synthesising diverse information and addressing complex research questions (Popay et al., 2006).

Advances in qualitative research methodology and foundational books have influenced the development of narrative synthesis (Thomas & Harden, 2008). A landmark book was Meta-Ethnography: Synthesising Qualitative Studies (1988) by Nancy Noblit and E G Peter Hare, which set out the main methodological analysis of synthesis via detailed, interpretively refined procedure. Subsequent elaborations and expansions – the theme synthesis of Thomas and Harden (2008), for instance, have enriched the methodological toolbox at researchers' disposal, and developments in digital technologies have made it easier to manage and analyse large volumes of qualitative data. (Popay et al., 2006; Thomas & Harden, 2008).

One notable advancement in the history of qualitative research synthesis techniques in healthcare settings is the Cochrane approach to narrative synthesis. The strategy was developed in 1993 by Cochrane et al. in response to the increasing demand for methodical techniques for combining qualitative data. Extending previous methodologies like theme analysis and meta-ethnography, the Cochrane method presented a methodical framework for amalgamating story data obtained from various qualitative investigations.

The iterative process of data extraction, processing, and thematic analysis, which seeks to uncover common threads and overarching themes across many narratives; is at the heart of this methodology. The Cochrane technique gives researchers a strong instrument for synthesising qualitative evidence and conducting high-quality qualitative research. Its transparent and systematic approach ensures clarity and rigour throughout the process. (Doyle, 2003; Drucker et al., 2016). The Cochrane approach further enriched narrative synthesis methods in reviews, through adopting 'evidence-based medicine' documentation, and hence, laid a foundation for the next generation of narrative synthesis also known as 'Realist and meta-narrative evidence synthesis.

1.2 Importance of Narrative Synthesis in Research

Narrative synthesis allows for the integration of diverse evidence: The integration of findings from diverse sources and studies with different methodologies, populations or interventions can provide a comprehensive understanding of a topic by considering a range of perspectives (Booth et al., 2019). Identifying gaps and inconsistencies in literature: by synthesising evidence from multiple studies, narrative synthesis can highlight gaps, inconsistencies, or contradictions in the literature. This can highlight the areas where additional research is required or where existing knowledge is insufficient or contradictory (Dixon-Woods et al., 2005).

Generating new insights: Narrative synthesis can help enable the generation of new hypotheses or theories, it achieves this by synthesising evidence from disparate sources and identifying emerging patterns and themes. These hypotheses can then be assessed in future research (Thomas & Harden, 2008).

Look for example at the following scenario:

Researchers conduct a narrative synthesis of studies on social media use and adolescent mental health, drawing from qualitative interviews, quantitative surveys, and longitudinal research. They identify patterns: some studies show a positive link between heavy social media use and mental health issues, while others report contradictory or inconclusive findings. From this synthesis, hypotheses emerge:

1- Diverse types and content of social media interactions may affect mental health outcomes differently.
2- Individual factors like personality traits and social support networks may moderate this relationship.

To evaluate these hypotheses, researchers plan experiments manipulating social media engagement types and analyse longitudinal data to understand how individual differences interact with social media over time. Narrative synthesis allows for making informed decisions: It gives a narrative overview of the evidence base, making it understandable to policymakers, practitioners, and other stakeholders who may lack statistical knowledge. This can help guide decision-making by offering a more nuanced view of the data and its consequences for practice or policy (Rodgers et al., 2009).

3- Complex phenomena that cannot be adequately expressed through simple statistical aggregation can be examined by exploring patterns, relationships, and contextual factors across studies (Erwin et al., 2011).

For example, let us assume that researchers are interested in understanding the efficacy of various interventions such as school-based programs, community initiatives, and parental involvement, in reducing childhood obesity. A simple statistical aggregation could entail pooling effect sizes from different studies to determine the overall average effectiveness of interventions. However, childhood obesity is a complex issue driven by a variety of factors including socioeconomic status, availability of nutritious foods, cultural norms, parental education, and community resources. As a result, a narrative synthesis method would entail delving into the qualitative features of the research, looking beyond impact estimates to comprehend the contextual factors and nuances associated with each intervention.

- **Various techniques of narrative synthesis**

There are different approaches for narrative synthesis, each can serve a defined purpose in meta-analysis reviews. In this part, two main types will be thoroughly highlighted in the context of general definitions and applications:

- Thematic synthesis
- Meta-narrative synthesis

2. THEMATIC SYNTHESIS

Thematic synthesis is a method used in the qualitative synthesis of primary research, characterised by identifying and analysing themes (reporting patterns) within a given data set. In addition, it also aims to draw an inclusive picture of the outcomes of a research question as well as its aspects-related interpretation. Three main stages comprise thematic synthesis which are (coding findings, descriptive themes, and analytical themes) (Dixon-Woods et al., 2005). Thematic synthesis possesses three advantages for researchers:

1- Allowing reviewers to line the expectations of a systematic review, besides having an 'audit trail' for the overall process interpretations and decisions
2- Enabling researchers as well as readers to closely follow conclusions related to the text in which the synthesis is based
3- Presenting a more formal approach to move from the text on pages all the way to synthesis

Thematic synthesis can be used to analyse any sets of data whether as small as two participants to studies with over 100 participants (Drucker et al., 2016). One of the hallmarks of thematic analysis is 'flexibility,' from exploring conclusions of psychological behaviour findings to interrogating patterns in a social study of certain populations, it has been proven to be 'critical' in establishing a framework in numerous studies, with a focus on psychological and social ones. Although flexible and useful, it cannot be concluded that thematic synthesis can be called an 'omni-method' in qualitative research, i.e. it has some shortcomings regarding other real-life clinical and basic science research.

2.1 Meta-narrative Synthesis

Meta-narrative synthesis is a part of narrative reviews that identifies and combines various research parts, and it involves searching for different data in unusual ways. It can be concluded that meta-narratives are composed of three main themes, the first one is an iterative configuration map, and aims to diversify traditions in each research area. Second, is an 'aggregative analysis/syntheses of various traditions in a map. The third one, is an iterative configuring in comparing of different meta-narratives to form a comprehensive integrative of an entire map (Gough, 2013).

Moreover, this approach is well known to have a 'summary signature' at the end of research where an overarching structured summary (reiterating) is written, rather than the typical concluding summary written in other forms of qualitative research. Constructing meta-narratives primarily relies on the following three steps:

1- Figuring out the hypothesis/phenomenon of interest: usually a trail of steps in order to find answers or fill a gap within research
2- Asking what is relevant: entailing a detailed narrow or broader way of searching for information, by using the following meta-sources (EBSCO, JSTOR, MEDLINE, PubMed, ScienceDirect, SCOPUS, Wiley)
3- Thorough reading and then re-reading: Researchers and reviewers should apply extensive cycles of reading and re-reading of relevant studies, this process can facilitate the data needed to be extracted and how they can be contextualised per each section of the review paper

2.2 Advantages of Narrative Synthesis

- **Flexibility**: Narrative synthesis allows for the integration of various study designs, procedures, and outcomes. This flexibility allows researchers to account for variation among studies (Sukhera, 2022).
- **Contextualization**: It provides a platform for contextualising findings within the larger research landscape, resulting in a more in-depth understanding of the topic under examination (Erwin et al., 2011).
- **Identification of patterns and themes**: Narrative synthesis enables researchers to identify patterns, themes, and linkages in data, revealing insights that quantitative analysis alone may not capture (Erwin et al., 2011).
- **Inclusion of qualitative evidence**: Narrative synthesis can combine qualitative and quantitative results to provide a more comprehensive understanding of complex phenomena (Major & Savin-Baden, 2010).
- **Accessibility**: It may be more accessible to researchers who lack sophisticated statistical competence, making it an effective tool for synthesising information in disciplines where statistical approaches are challenging to apply (Sukhera, 2022).
    - **Interpretation**: Narrative synthesis makes it easier to interpret findings by providing a narrative framework in which researchers can analyse the data' implications and potential real-world significance (Erwin et al., 2011; Major & Savin-Baden, 2010)

2.3 Challenges and Considerations

Addressing heterogeneity, managing bias, and balancing depth and breadth are essential challenges and considerations in narrative synthesis, as they require researchers to navigate variations in study designs, mitigate potential biases, and strike a balance between providing in-depth analysis while maintaining breadth of coverage across diverse sources of evidence (Lisy & Porritt, 2016)

Addressing heterogeneity

Recognize and address heterogeneity in the selected literature, including differences in study design, methodology, and conclusions.

Take this scenario for example: Suppose you are conducting a narrative synthesis of studies examining the effectiveness of different interventions for managing depression in adolescents. The studies in your review encompass various intervention types, such as cognitive-behavioural therapy (CBT), medication, mindfulness-based interventions, and exercise programs. Additionally, the studies involve diverse populations, including adolescents from different socio-economic backgrounds and cultural contexts.

Steps to consider when addressing heterogeneity (Barnett-Page & Thomas, 2009; Hong & Pluye, 2018; Schroll et al., 2011).

Clearly define your inclusion criteria for selecting studies to be involved in the narrative synthesis, consider factors such as study design, population characteristics, and intervention types, and measure outcomes

- Conduct a comprehensive search of relevant literature using multiple databases and sources to identify studies meeting your inclusion criteria.
- Create a systematic approach to extract relevant data from each included study, extract information on study characteristics, participant demographics, intervention type, results and significant findings established.
- To assess study heterogeneity amongst the included studies, examine the differences in study designs, participant characteristics, intervention components, outcome measures and study venues.
- Identify Patterns and Themes: Look for patterns and themes emerging from the data extracted from the included studies. Identify commonalities and differences in intervention effectiveness across different populations and contexts.
- Consider subgroup analysis and explore how the effectiveness of interventions may vary across different subpopulations or intervention types, for example, you may analyse whether certain interven-

tions are more effective amongst specific age groups or individuals from certain socio-economic backgrounds.

- Lastly describe the key characteristics of each study, summarising the interventions and outcomes, and discussing similarities and differences among the studies. Address any conflicting evidence or contradictory findings by critically evaluating the methodological quality of the studies and considering potential sources of bias or confounding factors.
- From this you can draw conclusions about the effectiveness of interventions for managing depression in adolescents while acknowledging the heterogeneity across studies.

Managing Bias: Be mindful of potential biases in the synthesis process, such as publication bias or researcher bias.

- The risk of bias in narrative reviews can be managed by developing a comprehensive search strategy to identify relevant literature from multiple sources, using a combination of electronic databases, grey literature, hand-searching, and consultation with experts to minimise the risk of publication bias (Drucker et al., 2016)
- Ensure clear inclusion and exclusion criteria based on research questions or objectives. Be explicit about the criteria used to select studies to ensure that all relevant literature is considered while minimising the risk of selection bias (Drucker et al., 2016).
- Conduct a critical appraisal of the included studies to assess their methodological quality and risk of bias. Use appropriate tools or frameworks to evaluate study design, sample size, outcome measures, and potential sources of bias (Hong & Pluye, 2018).
- Balancing Depth and Breadth: Strike a balance between providing sufficient depth of analysis and breadth of coverage in the narrative synthesis. Focus on key themes and findings while also acknowledging the broader context and complexities of the topic.

2.4 Limitations of Narrative Synthesis

- Subjectivity: The interpretation of findings in narrative synthesis is fundamentally subjective because it relies on the researchers' judgement to detect patterns and themes. This subjectivity may cause bias in the synthesis process (Barnett-Page & Thomas, 2009; Rodgers et al., 2009).
- Lack of transparency: Unlike quantitative methods such as meta-analysis, narrative synthesis may lack transparency in terms of the procedures employed to select, extract, and synthesise data, potentially jeopardising the reproducibility of results (Campbell et al., 2019).
- Difficulty with quantification: Narrative synthesis does not produce quantitative estimates of effect sizes or precision measures, limiting its ability to quantify the degree of correlations between variables or evidence strength (Barnett-Page & Thomas, 2009; Rodgers et al., 2009).
- Researcher bias: Researchers may unintentionally or purposely choose studies that fit their preexisting views or hypotheses, resulting in biassed conclusions in the synthesis (Pae, 2015).
- Difficulty with heterogeneity: Narrative synthesis may struggle to oversee variability between research, especially when there are significant differences in study designs, demographics, or interventions (Campbell et al., 2019).
- Limited generalizability: Because narrative synthesis is qualitative, the findings may be less generalizable than quantitative methods, which can provide more precise estimates of population effects (Barnett-Page & Thomas, 2009; Rodgers et al., 2009).

Narrative synthesis is a powerful tool, but like any research method, it is vulnerable to bias as previously mentioned. Here's how researchers can be detectives and sniff out potential problems:

- **Picking the Right Studies:** Imagine a researcher only looking at published papers, missing unpublished research (selection bias). To avoid this, they can search widely across databases and include unpublished studies (grey literature) (Page et al., 2021). Clear rules for what studies to include and exclude (inclusion/exclusion criteria) also help. Tools like PRISMA flow diagrams keep track of the selection process for transparency.

- **Published vs. Real World:** Sometimes, studies with positive results are more likely to get published, skewing the overall picture (publication bias). Researchers can use tools like funnel plots and statistical tests to check for this.
- **Reporting the Good findings:** Imagine a researcher only reporting findings that support their initial idea (reporting bias). To catch this, researchers can compare study plans with published results. Tools like registries that store study plans can help identify discrepancies.
- **Confirmation bias:** Researchers might favour evidence that confirms their existing beliefs (confirmation bias). Using blind review processes, where reviewers do not know the study's hypothesis, can help reduce this.

Here's how researchers can mitigate the negative influence of bias:

- **Transparency is Key:** Following clear reporting guidelines like the Preferred Reporting Items for Systematic Reviews and Meta-Analyses (PRISMA) ensures a clear and unbiased picture of the research process.
- **Assessing the Strength:** Running sensitivity analyses helps see how changing the chosen studies or quality standards affects the results, revealing potential bias.
- **Getting Different Perspectives:** Involving experts and end-users in the research process (stakeholder involvement) brings in diverse viewpoints and reduces bias. Methods like the Delphi technique and consensus conferences are helpful tools for this.
- **Quality Matters:** Assessing the quality of included studies with established tools helps weed out low-quality studies that could skew the results.
- **Seeing the Bigger Picture:** Using multiple data sources and combining narrative synthesis with other methods (like meta-analysis)

provides a more complete picture and reduces the impact of bias (Sterne et al., 2011).

- **Sharpening the Skills:** Training researchers on identifying and mitigating bias ensures an elevated level of awareness and skill in managing these challenges.

Recent research emphasizes the importance of being transparent and following rigorous methods to reduce bias. New tools and updated guidelines are constantly being developed to ensure the validity of narrative synthesis. By being detectives of bias, researchers can ensure their narrative synthesis is dependable and leads to trustworthy conclusions (Munn et al., 2020).

2.5 Summary of the Advantages and Limitations

Table 1. Advantages and Limitations of narrative synthesis

Advantages	Disadvantages
Narrative synthesis can facilitate different research topics and their relevant discussions.	Inconsistent transparencies are often noted in narrative synthesis, as it has been reported that fewer than 30% of all reviews reflected appropriate measures for robust transparency in synthesis (Campbell et al., 2019).
it offers outstanding flexibility in showing data, reporting intervention results, and defining the used methodologies	One notable weakness is that sometimes narrative reviews can be biassed, according to the author's beliefs/intentions.
It enables exploring a wide range of techniques and approaches in answering specific research questions.	Ownership issues, for instance when a researcher wants to narrate a particular concept that differs from the general theme (agreed upon) of the review also known as, 're-storying' (Josselson, 2007).
Narrative synthesis is considered a 'reliable' form of qualitative research and data reporting, moreover, it is one of the most widely used academic writing methods for researchers across various fields. Those in healthcare and translational sciences.	

3.COUPLE OF REAL-WORLD EXAMPLES HIGHLIGHTING NARRATIVE SYNTHESIS IMPORTANCE AND USABILITY

- Understanding Cancer Care Beyond Numbers: for instance, if we thought about a researcher wanting to grasp the full picture of cancer patient's experience with a certain type of immunotherapy. Numbers solely, might tell us the average survival rate, however narrative synthesis goes deeper an takes the extra mile. By allowing the researcher to analyse stories and experiences from different patients, including their fears, hopes, adverse events and overall quality of life. Through combining all these narratives together, researchers can draw a richer and more comprehensive image, including the emotional and psychological aspect of patients. This can further inform adequate interventions needed to not only address the physical side of disease, but also the overall patient experience.
- Community Policing: If we suppose that a researcher wants to understand how effective community policing can curb the rate of crime. A classical approach may merely look at the average crime reduction rate across various programs. Whereas real-world situations are often more complicated and in an informal description "messy." On the other hand, narrative synthesis can allow a deeper look into the root cause and consequences of a certain phenomenon, in our example, it can help researchers to better interact with residents, types of crimes targeted and how well the community responds to the policing program. The richer story reveals not only "did it work?" but also "how" and under what circumstances. This can present valuable insights for public policymakers to craft effective community policing strategies.

These are just a couple of examples on how narrative synthesis goes beyond just "cold facts." Through combining qualitative and inter-mixed studies, which help us to understand human stories behind research. The next section will demonstrate in detail several case studies, showing how narrative synthesis has contributed to real-life scenarios.

Case Study 1: Evaluating Chronic Pain Interventions

Background

Chronic pain management presents a persistent challenge within the healthcare landscape. While a multitude of interventions have been proposed and investigated, their efficacy often demonstrates significant variability. To address this, a narrative synthesis was employed to comprehensively evaluate the effectiveness of various chronic pain interventions, aiming to provide a more nuanced understanding of the existing evidence base. This approach facilitates identification of recurring themes and potential moderators of treatment response, ultimately informing the development of more targeted and effective chronic pain management strategies.

Methodology

The narrative synthesis incorporated both qualitative and quantitative data, drawing upon a rigorous literature search conducted across a variety of databases. Studies were meticulously selected according to predetermined inclusion criteria, prioritizing randomized controlled trials (RCTs), cohort studies, and qualitative research that delved into patient experiences with chronic pain management interventions.

Findings

The narrative synthesis revealed the following insights:

1- Chronic Pain Management: A Multifaceted Approach

The narrative synthesis revealed a complex landscape of chronic pain interventions, encompassing pharmacological treatments, cognitive-behavioural therapy (CBT), and physical exercise. Notably, the success of each intervention hinged on the specific patient population and the type of pain they experienced.

2- Patient-Cantered Care is Paramount

Qualitative studies underscored the critical role of personalized treatment plans that acknowledge patient preferences and experiences. For instance, research by Eccleston et al. (2017) demonstrated the enhanced efficacy of CBT when tailored to individual needs (Eccleston et al., 2017).

3- Synergy in Combination Therapies

The synthesis highlighted the potential of combination therapies, suggesting they often yield superior outcomes compared to single interventions. Studies by Geneen et al. (2017) suggests that combining pharmacological treatments with physical and psychological therapies holds promise for significantly improving overall patient well-being, as well as reflecting a positive outcome on the overall quality of life (Geneen et al., 2017).

Conclusion: A Multifaceted Path Forward

By providing a comprehensive overview of chronic pain interventions, the narrative synthesis underscores the importance of a multifaceted, patient-cantered approach. This case study reinforces the value of this methodology in integrating diverse evidence and offering actionable insights for effective clinical practice in chronic pain management.

Case Study 2: Mental Health Interventions in Refugees Populations

Background

Refugee Populations and Mental Health: Displacement, trauma, and the stressors of resettlement pose unique mental health challenges for refugees. To evaluate the effectiveness of interventions aimed at this vulnerable population, researchers employed a narrative synthesis approach.

Methodology

The synthesis encompassed studies from diverse contexts and countries. It incorporated various research designs, including randomized controlled trials (RCTs), observational studies, and qualitative research exploring refugee experiences. A comprehensive search strategy targeted major databases like PubMed, PsycINFO, and EMBASE, along with unpublished research.

Findings

Critical Themes: The synthesis identified several crucial themes:

- **Culturally Tailored Interventions:** Interventions demonstrating cultural sensitivity towards the specific refugee population were more effective. Programs that included cultural liaisons or adapted materials to the local context yielded improved engagement and outcomes (Bentley et al., 2019).
- **Community-Based Support Systems:** Community-based interventions, such as group therapy or peer support programs, proved particularly beneficial. This approach facilitated the reduction of stigma and provided essential social support for mental health recovery (Tol et al., 2011).
- **Addressing Access Barriers:** Common barriers to accessing mental health services included language difficulties, mistrust of healthcare providers, and logistical challenges. Addressing these barriers through translation services, trust-building initiatives, and accessible service locations enhanced intervention uptake (Purgato et al., 2018).

Conclusion

This case study underscores the significance of culturally sensitive, community-based interventions in improving mental health outcomes for refugees. Narrative synthesis facilitated the integration of diverse research methodologies and findings, offering a nuanced understanding of effective intervention strategies for this population.

Case Study 3: School-Based Obesity Prevention Programs

Background

Childhood obesity represents a growing public health concern. School-based programs are frequently implemented to prevent obesity; however, their effectiveness varies. A narrative synthesis was conducted to assess the success of these programs and identify factors contributing to their efficacy.

Methodology

The synthesis included various research designs: RCTs, quasi-experimental studies, and qualitative research. The search strategy encompassed electronic databases, unpublished research (grey literature), and manual searches of reference lists from other inclusive studies. Standardized tools were as well employed to evaluate the quality incorporated research.

Findings

Key Insights: The synthesis revealed several key insights:

- **Multicomponent Programs:** Programs that combined elements like nutritional education, physical activity promotion, and behavioural modification strategies demonstrated greater effectiveness in reducing obesity rates. Standalone interventions, such as solely increasing physical activity, yielded limited success (Waters et al., 2011).
- **Parental Engagement:** Interventions that involved parents, either through educational sessions or take-home materials, achieved better outcomes. Parental engagement reinforced healthy behaviours at

home, complementing school-based efforts (González-Suárez et al., 2013).

- **Long-Term Follow-Up:** Programs with long-term follow-up demonstrated sustained benefits compared to those with short-term follow-up. Continued support and reinforcement of healthy habits were critical for maintaining positive outcomes (Langford et al., 2014).

Conclusion

The narrative synthesis highlighted that successful school-based obesity prevention programs are multifaceted, incorporate parental engagement, and include long-term follow-up. This comprehensive approach provides a clearer understanding of what constitutes effective childhood obesity prevention strategies and the underlying mechanisms of their success.

Case Study 4: Palliative Care for Advanced Cancer Cases

Background

Palliative care plays a vital role in improving the quality of life for patients with advanced cancer. However, the effectiveness of various palliative care models can differ significantly. To evaluate these models and identify best practices, a narrative synthesis was conducted.

Methodology

The synthesis incorporated a range of study designs, including randomized controlled trials (RCTs), cohort studies, and qualitative research. The search strategy encompassed multiple databases, grey literature, and consultations with experts in the field of palliative care.

Findings

The synthesis revealed several key themes:

Holistic Care Models: Models that addressed the physical, emotional, social, and spiritual needs of patients demonstrated greater effectiveness in improving their quality of life. These holistic approaches often involved multidisciplinary teams, bringing together various specialists (Kavalieratos et al., 2016).

Early Integration: Early integration of palliative care alongside standard oncological care resulted in better patient outcomes compared to late-stage integration. Early intervention facilitated more effective symptom management and provided crucial psychological support (Temel et al., 2010).

Communication and Decision-Making: Effective communication between healthcare providers, patients, and families proved to be essential. Clear discussions about goals of care and patient preferences improved patient satisfaction and reduced unnecessary interventions (Smith et al., 2014).

Conclusion

This case study underscores the value of early, holistic, and well-communicated palliative care models for advanced cancer patients. Narrative synthesis facilitated the integration of diverse evidence, providing a comprehensive view of effective practices in palliative care.

Case Study 5: Bridging the Gap: E-Health in Rural Healthcare

Background

Limited access to healthcare services is a persistent challenge for rural communities. E-health solutions, such as telemedicine and mobile health apps, offer potential benefits in this context. A narrative synthesis was conducted to evaluate the implementation and effectiveness of these solutions in rural settings.

Methodology

The synthesis incorporated quantitative and qualitative studies from diverse rural contexts. The search strategy encompassed electronic databases, grey literature, and consultations with e-health experts. Standardized tools were employed to assess the quality of the included studies.

Findings

Key Findings: The synthesis revealed several key findings:

- **Improved Access and Utilization:** E-health solutions demonstrably improved access to healthcare services for rural populations. Telemedicine, in particular, reduced travel time and costs, facilitating patients' ability to receive timely care (Shaw et al., 2018).
- **Technological Barriers:** Common barriers included limited internet connectivity, lack of digital literacy, and resistance to adopting innovative technologies. Addressing these barriers through infrastructure improvements and patient education was essential for successful implementation (Gagnon et al., 2016).
- **Integration with Existing Services:** E-health solutions were most effective when integrated with existing healthcare services. Combining telemedicine with in-person visits, when necessary, provided a more comprehensive care model.

Case Study 6: Mental Health Interventions for Teenagers

Background

Teenagers' mental health is a critical public health concern, with increasing rates of depression, anxiety, and other mental health disorders. A narrative synthesis was conducted to evaluate various mental health interventions for adolescents, focusing on effectiveness and implementation challenges.

Methodology

The synthesis included a wide range of studies, from randomized controlled trials (RCTs) to qualitative research, covering several types of interventions such as school-based programs, online therapy, and community-based initiatives. The search strategy was extensive, encompassing numerous databases and grey literature.

Findings

Key Themes: The synthesis identified several crucial themes:

- **School-Based Programs:** These programs were effective in reducing symptoms of depression and anxiety, particularly when they incorporated cognitive-behavioural components and were delivered by trained professionals (Werner-Seidler et al., 2017).
- **Digital Interventions:** Online therapy and mobile apps showed promise, especially for adolescents who preferred the anonymity and convenience of digital platforms. However, engagement and adherence remained challenges that needed to be addressed (Clarke et al., 2015).
- **Community-Based Approaches:** Programs that included community resources, such as sports clubs and youth centres, were effective in promoting mental well-being. Such interventions often relied on peer support and mentoring (Loades et al., 2020).

Case Study 7: Rehabilitation Programs for Stroke Survivors

Background

Stroke rehabilitation is crucial for recovery and quality of life improvement in stroke survivors. A narrative synthesis was conducted to evaluate different rehabilitation programs and their effectiveness.

Methodology

The synthesis included a mix of RCTs, cohort studies, and qualitative research. The search strategy was broad, covering electronic databases, grey literature, and manual searches of reference lists.

Findings

Key insights from the synthesis included:

- **Multidisciplinary Rehabilitation**: Programs involving multidisciplinary teams, including physiotherapists, occupational therapists, and speech therapists, were most beneficial in improving functional outcomes (Langhorne et al., 2011).
- **Home-Based Rehabilitation**: Home-based programs were as effective as hospital-based ones, provided they included regular follow-up and support. These programs were particularly beneficial for patients with mobility problems (Laver et al., 2020).
- **Patient and Family Involvement**: Involving patients with their families in the rehabilitation process has enhanced adherence and outcomes. Family members often provided crucial support and motivation (Galvin et al., 2011).

Conclusion

The narrative synthesis highlighted the effectiveness of multidisciplinary and home-based rehabilitation programs, as well as the importance of patient and family involvement. This case study illustrates the utility of narrative synthesis in evaluating complex healthcare interventions.

These case studies demonstrate the diverse applications and benefits of narrative synthesis in various research contexts. From chronic pain management to mental health interventions and stroke rehabilitation, narrative synthesis enables the integration of heterogeneous evidence, providing comprehensive insights that inform practice and policy. By highlighting key themes, identifying effective strategies, and addressing barriers, narrative synthesis serves as a valuable tool for researchers and healthcare practitioners alike.

4. PRACTICAL GUIDE ON HOW TO CONSTRUCT THE MAIN ELEMENTS OF NARRATIVE SYNTHESIS

At first, you must:

- Identify the research question
  - To develop a theory and see how a certain intervention works, why and for whom
  - To consider a scheme of how this intervention works and how it is affected by several factors and variables
  - To make a general outline for your proposed synthesis
- Literature synthesis and finding selection
  - To conduct thorough research about the research question
  - To select the most relevant, appropriate, and unbiased studies connected to the research question and proposed interventions
  - To group the studies by common factors such as: comparison, format, and complexity
  - To document primary and secondary events happening and to tabulate the results/data
  - To transform the data, for example to be either statistical or descriptive style
  - To translate and interpret the data through content analysis or thematic format which is commonly used in narrative synthesis in general

Data Sorting and Extrapolation

- To have a systematic look about within and between studies to discover differences in patterns and the efficiency levels of interventions as:

  1- Within studies: to assess different effects of an intervention between populations/groups or in various settings

2- Across studies: for instance, randomised controlled trials (RCTs) may rule out the lack of effect of a particular intervention whereas, other weaker types of studies (quasi-RCTs) may find some patterns and connections based on differences within populations such as (adult vs young)

- To consider whether the following points can elaborate on differences between studies

  1- Differences in outcomes

  2- Variability in study designs

  3- Population/settings variation

- To examine the effects of heterogeneity in synthesis construction, i.e. to look for methodological differences across studies and how their respective results are presented in reviews

Synthesis Construction and Interpretation

- To synthesise your outcomes/findings into a coherent narrative style
- To assess the overall strength of evidence
- To screen for any inconsistency patterns
- To set a solid conclusion for the main findings

Reporting and Final Assessment of the Overall Synthesis Strength

- To assess the quality and quantity of evidence
- To minimise any pertaining biases
- To present your final piece of work in a transparent and clear academic manner; written report/manuscript

As with any element in academic research, narrative synthesis has its advantages and limitations as well. Across various fields, there is still no definite consensus about solid shortcomings for narrative synthesis; however, academics have drawn some lines on common themes about narrative synthesis. The differences between distinct types of narrative reviews styles will be shown in the table.2, utilizing a real-life example of team of detectives on a mission, for easier illustration for students and researchers across board of disciplines.

Table 2. Simplified summary of comparative analysis between narrative synthesis and other syntheses

Method	Such as	Good for	Considerations and challenges
Narrative synthesis	A research team on a detective-style mission with all the clues	Comprehending complex stories and findings of a research	Bias and not being capable of directly comparing studies
Meta-analysis	Gathering and ordering all witness statements	Having a clear picture of a certain effect's influence	Studies needing to be unified on using the same parameters and measurements
Thematic analysis	Finding patterns and common threads across different interviews	Identifying crucial and core ideas that emerge from research	Losing details, shifting focus when grouping themes
Meta-narrative synthesis	Evaluating how different witness testimonies and accounts can fit together	Disclosing the bigger story that different studies tell	Requiring adequate and professional skills to analyse

Always Remember (Take Home Message):

1- Narrative synthesis: Is great for exploring the "why" and "how" behind any research. However, one should be mindful of their own perspective.
2- Meta-analysis: Perfect and suitable for getting a precise answer, but it works only for studies speaking the same language (data-wise).
3- Thematic analysis: Useful for spotting and grouping common threads across broader research.
4- Meta-narrative synthesis: Can aid you to see the bigger picture emerging from myriad studies.

5. CONCLUDING REMARKS

In summary, narrative synthesis is a valuable methodological tool for researchers, offering researchers a systematic way to collate, interpret, and integrate evidence from diverse sources. While it has limitations such as subjectivity and lack of standardised guidelines, it remains a valuable method for integrating qualitative research findings. Moreover, narrative synthesis is still considered the gold standard for flexible data presentation, facilitating finding answers to multifaceted issue issues in research. Despite its utility, researchers must remain cognizant of its limitations and strive to maintain transparency and rigour throughout the synthesis process, because while narrative synthesis offers flexibility in data presentation and facilitates the exploration of multifaceted research issues, it is not without drawbacks such as subjectivity and the absence of standardised guidelines. Nevertheless, despite its inherent challenges, narrative synthesis remains a prominent approach for integrating qualitative research findings, effectively guiding policy and practice while advancing knowledge across various fields of study when executed carefully and transparently. Lastly, the chapter highlighted crucial case studies, where narrative synthesis can be relatable in solving complex research questions.

REFERENCES

Barnett-Page, E., & Thomas, J. (2009). Methods for the synthesis of qualitative research: A critical review. *BMC Medical Research Methodology*, *9*(1), 59. Advance online publication. DOI: 10.1186/1471-2288-9-59 PMID: 19671152

Bentley, J. A., Thoburn, J. W., & Stewart, D. (2019). Cultural adaptations of mental health interventions for refugee populations. *The Journal of Nervous and Mental Disease*, *207*(12), 1044–1052. DOI: 10.1097/NMD.0000000000001092

Booth, A., Moore, G., Flemming, K., Garside, R., Rollins, N., Tunçalp, Ö., & Noyes, J. (2019). Taking account of context in systematic reviews and guidelines considering a complexity perspective. *BMJ Global Health*, *4*(Suppl 1), e000840. DOI: 10.1136/bmjgh-2018-000840 PMID: 30775011

Campbell, M., Katikireddi, S. V., Sowden, A., & Thomson, H. (2019). Lack of transparency in reporting narrative synthesis of quantitative data: A methodological assessment of systematic reviews. *Journal of Clinical Epidemiology*, *105*, 1–9. DOI: 10.1016/j.jclinepi.2018.08.019 PMID: 30196129

Clarke, A. M., Kuosmanen, T., & Barry, M. M. (2015). A systematic review of online youth mental health promotion and prevention interventions. *Journal of Youth and Adolescence*, *44*(1), 90–113. DOI: 10.1007/s10964-014-0165-0 PMID: 25115460

Dixon-Woods, M., Agarwal, S., Jones, D., Young, B., & Sutton, A. (2005). Synthesising qualitative and quantitative evidence: A review of possible methods. *Journal of Health Services Research & Policy*, *10*(1), 45–53. DOI: 10.1177/135581960501000110 PMID: 15667704

Doyle, L. H. (2003). Synthesis through meta-ethnography: Paradoxes, enhancements, and possibilities. *Qualitative Research*, *3*(3), 321–344. DOI: 10.1177/1468794103033003

Drucker, A. M., Fleming, P., & Chan, A.-W. (2016). Research techniques made simple: Assessing risk of bias in systematic reviews. *The Journal of Investigative Dermatology, 136*(11), e109–e114. DOI: 10.1016/j.jid.2016.08.021 PMID: 27772550

Eccleston, C., Fisher, E., & Vowles, K. E.. (2017). Psychological therapies (Internet-delivered) for the management of chronic pain in adults. *Cochrane Database of Systematic Reviews*, 2. PMID: 24574082

Erwin, E. J., Brotherson, M. J., & Summers, J. A. (2011). Understanding qualitative metasynthesis. *Journal of Early Intervention, 33*(3), 186–200. DOI: 10.1177/1053815111425493

Gagnon, M.-P., Desmartis, M., & Labrecque, M.. (2016). Implementation of telehealth in an interprofessional rehabilitation university clinic: A mixed-method study. *Telemedicine Journal and e-Health, 22*(11), 951–959. DOI: 10.1089/tmj.2015.0221

Galvin, R., Cusack, T., & Stokes, E. (2011). To what extent are family members and friends involved in physiotherapy and the delivery of exercises to people with stroke? A qualitative study. *Disability and Rehabilitation, 33*(15–16), 1395–1404. DOI: 10.3109/09638288.2010.533244

Geneen, L. J., Moore, R. A., & Clarke, C.. (2017). Physical activity and exercise for chronic pain in adults: An overview of Cochrane reviews. *Cochrane Database of Systematic Reviews*, 4.

González-Suárez, C., Worley, A., Grimmer-Somers, K., & Dones, V. (2013). School-based interventions on childhood obesity: A meta-analysis. *American Journal of Preventive Medicine, 44*(5), 517–523. DOI: 10.1016/j.amepre.2012.12.007 PMID: 19840696

Gough, D. (2013). Meta-narrative and realist reviews: Guidance, rules, publication standards and quality appraisal. *BMC Medicine, 11*(1), 22. Advance online publication. DOI: 10.1186/1741-7015-11-22 PMID: 23360691

Hannes, K., & Macaitis, K. (2012). A move to more systematic and transparent approaches in qualitative evidence synthesis: Update on a review of published papers. *Qualitative Research, 12*(4), 402–422. DOI: 10.1177/1468794111432992

Hong, Q. N., & Pluye, P. (2018). A conceptual framework for critical appraisal in systematic mixed studies reviews. *Journal of Mixed Methods Research, 13*(4), 1558689818770058. Advance online publication. DOI: 10.1177/1558689818770058

. Josselson, R. (2007). The ethical attitude in narrative research. *Principles of Qualitative Research.*

Kavalieratos, D., Corbelli, J., Zhang, D., Dionne-Odom, J. N., Ernecoff, N. C., Hanmer, J., Hoydich, Z. P., Ikejiani, D. Z., Klein-Fedyshin, M., Zimmermann, C., Morton, S. C., Arnold, R. M., Heller, L., & Schenker, Y. (2016). Association between palliative care and patient and caregiver outcomes: A systematic review and meta-analysis. *Journal of the American Medical Association, 316*(20), 2104–2114. DOI: 10.1001/jama.2016.16840 PMID: 27893131

Langford, R., Bonell, C. P., Jones, H. E., Pouliou, T., Murphy, S. M., Waters, E., Komro, K. A., Gibbs, L. F., Magnus, D., & Campbell, R. (2014). The WHO Health Promoting School framework for improving the health and well-being of students and their academic achievement. *Cochrane Database of Systematic Reviews, 2014*(4), 4. DOI: 10.1002/14651858. CD008958.pub2 PMID: 24737131

Langhorne, P., Bernhardt, J., & Kwakkel, G. (2011). Stroke rehabilitation. *Lancet, 377*(9778), 1693–1702. DOI: 10.1016/S0140-6736(11)60325-5 PMID: 21571152

Laver, K., Lange, B., & George, S.. (2020). Virtual reality for stroke rehabilitation. *Cochrane Database of Systematic Reviews*, 11.

Lisy, K., & Porritt, K. (2016). Narrative synthesis. *International Journal of Evidence-Based Healthcare, 14*(4), 201. DOI: 10.1097/01. XEB.0000511348.97198.8c

Loades, M. E., Chatburn, E., Higson-Sweeney, N., Reynolds, S., Shafran, R., Brigden, A., Linney, C., McManus, M. N., Borwick, C., & Crawley, E. (2020). Rapid systematic review: The impact of social isolation and loneliness on the mental health of children and adolescents in the context of COVID-19. *Journal of the American Academy of Child and Adolescent Psychiatry*, *59*(11), 1218–1239.e3. DOI: 10.1016/j.jaac.2020.05.009 PMID: 32504808

Major, C. H., & Savin-Baden, M. (2010). *An introduction to qualitative research synthesis*. Routledge., DOI: 10.4324/9780203497555

Munn, Z., Moola, S., Lisy, K., Riitano, D., & Tufanaru, C. (2020). Methodological guidance for systematic reviews of observational epidemiological studies reporting prevalence and cumulative incidence data. *International Journal of Evidence-Based Healthcare*, *18*(3), 147–153. DOI: 10.1097/XEB.0000000000000054 PMID: 26317388

Noyes, J., Lewin, S., & Rees, R. (2011). *A guide to synthesising qualitative research for researchers undertaking health technology assessments and systematic reviews*. University of Stirling. https://www.stir.ac.uk/research/hub/publication/817639

Pae, C. U. (2015). Why systematic review rather than narrative review? *Psychiatry Investigation*, *12*(3), 417. DOI: 10.4306/pi.2015.12.3.417 PMID: 26207140

Page, M. J., McKenzie, J. E., Bossuyt, P. M., Boutron, I., Hoffmann, T. C., Mulrow, C. D., Shamseer, L., Tetzlaff, J. M., Akl, E. A., Brennan, S. E., Chou, R., Glanville, J., Grimshaw, J. M., Hróbjartsson, A., Lalu, M. M., Li, T., Loder, E. W., Mayo-Wilson, E., McDonald, S., & Moher, D. (2021). The PRISMA 2020 statement: An updated guideline for reporting systematic reviews. *BMJ (Clinical Research Ed.)*, *372*(71), n71. Advance online publication. DOI: 10.1136/bmj.n71 PMID: 33782057

Popay, J., Roberts, H., & Sowden, A.. (2006). *Guidance on the conduct of narrative synthesis in systematic reviews*. ESRC Methods Programme.

Purgato, M., Tol, W. A., & Bass, J. K.. (2018). Psychosocial interventions for children exposed to traumatic events in low- and middle-income countries: A systematic review and meta-analysis. *Journal of Child Psychology and Psychiatry, and Allied Disciplines*, *59*(7), 767–783. DOI: 10.1111/jcpp.12851

Rodgers, M., Sowden, A., Petticrew, M., Arai, L., Roberts, H., Britten, N., & Popay, J. (2009). Testing methodological guidance on the conduct of narrative synthesis in systematic reviews. *Evaluation*, *15*(1), 49–73. DOI: 10.1177/1356389008097871

Sattar, R., Lawton, R., Panagioti, M., & Johnson, J. (2021). Meta-ethnography in healthcare research: A guide to using a meta-ethnographic approach for literature synthesis. *BMC Health Services Research*, *21*(1), 50. Advance online publication. DOI: 10.1186/s12913-020-06049-w PMID: 33419430

Schroll, J. B., Moustgaard, R., & Gøtzsche, P. C. (2011). Dealing with substantial heterogeneity in Cochrane reviews: Cross-sectional study. *BMC Medical Research Methodology*, *11*(1), 22. Advance online publication. DOI: 10.1186/1471-2288-11-22 PMID: 21349195

Shaw, E. K., Howard, J., & West, D. R.. (2018). The role of the practice facilitator in implementing the patient-centered medical home (PCMH). *Journal of the American Board of Family Medicine : JABFM*, *31*(5), 762–773. DOI: 10.3122/jabfm.2018.05.170294

Smith, S., Brick, A., O'Hara, S., & Normand, C. (2014). Evidence on the cost and cost-effectiveness of palliative care: A literature review. *Palliative Medicine*, *28*(2), 130–150. DOI: 10.1177/0269216313493466 PMID: 23838378

Sterne, J. A. C., Sutton, A. J., Ioannidis, J. P. A., Terrin, N., Jones, D. R., Lau, J., Carpenter, J., Rucker, G., Harbord, R. M., Schmid, C. H., Tetzlaff, J., Deeks, J. J., Peters, J., Macaskill, P., Schwarzer, G., Duval, S., Altman, D. G., Moher, D., & Higgins, J. P. T. (2011). Recommendations for examining and interpreting funnel plot asymmetry in meta-analyses of randomised controlled trials. *BMJ (Clinical Research Ed.)*, *343*(jul22 1), d4002. DOI: 10.1136/bmj.d4002 PMID: 21784880

Sukhera, J. (2022). Narrative reviews: Flexible, rigorous, and practical. *Journal of Graduate Medical Education*, *14*(4), 414–417. DOI: 10.4300/JGME-D-22-00480.1 PMID: 35991099

Temel, J. S., Greer, J. A., Muzikansky, A., Gallagher, E. R., Admane, S., Jackson, V. A., Dahlin, C. M., Blinderman, C. D., Jacobsen, J., Pirl, W. F., Billings, J. A., & Lynch, T. J. (2010). Early palliative care for patients with metastatic non-small-cell lung cancer. *The New England Journal of Medicine*, *363*(8), 733–742. DOI: 10.1056/NEJMoa1000678 PMID: 20818875

Thomas, J., & Harden, A. (2008). Methods for the thematic synthesis of qualitative research in systematic reviews. *BMC Medical Research Methodology*, *8*(1), 45. DOI: 10.1186/1471-2288-8-45 PMID: 18616818

Tol, W. A., Barbui, C., Galappatti, A., Silove, D., Betancourt, T. S., Souza, R., Golaz, A., & van Ommeren, M. (2011). Mental health and psychosocial support in humanitarian settings: Linking practice and research. *Lancet*, *378*(9802), 2126–2132. DOI: 10.1016/S0140-6736(11)61094-5 PMID: 22008428

Waters, E., de Silva-Sanigorski, A., & Hall, B. J.. (2011). Interventions for preventing obesity in children. *Cochrane Database of Systematic Reviews*, 12. PMID: 22161367

Werner-Seidler, A., Perry, Y., Calear, A. L., Newby, J. M., & Christensen, H. (2017). School-based depression and anxiety prevention programs for young people: A systematic review and meta-analysis. *Clinical Psychology Review*, *51*, 30–47. DOI: 10.1016/j.cpr.2016.10.005 PMID: 27821267

Wong, G., Greenhalgh, T., Westhorp, G., Buckingham, J., & Pawson, R. (2013). RAMESES publication standards: Meta-narrative reviews. *BMC Medicine*, *11*(1), 20. Advance online publication. DOI: 10.1186/1741-7015-11-20 PMID: 23360661

APPENDIX

Additional Resources

- Testing Methodological Guidance on the Conduct of Narrative Synthesis in Systematic Reviews. https://www.lancaster.ac.uk/media/lancaster-university/content-assets/documents/fhm/dhr/chir/Rodgersetal.pdf
- Cochrane Consumers and Communication (Data Synthesis and Data Analysis). http://cccrg.cochrane.org
- The Oxford Handbook of the Science of Science Communication, chapter 9 (Publication Bias in Science: What Is it? Why Is it problematic? And How Can It Be Addressed?). https://doi.org/10.1093/oxfordhb/9780190497620.013.10

Chapter 2
Exploring Scoping and Mapping Reviews

Sai Kaushik Upputuru
https://orcid.org/0009-0007-1899-5025
University of South Florida, USA

Cameron Sabet
Department of Medicine, School of Medicine, Georgetown University, USA

Trisha Bhathivi
https://orcid.org/0009-0002-8863-3402
College of Arts and Sciences, University of South Florida, USA

Ramez M. Odat
https://orcid.org/0009-0008-3423-4055
Faculty of Medicine, Jordan University of Science and Technology, Irbid, Jordan

Dang Nguyen
Judy Genshaft Honors College, University of South Florida, USA

DOI: 10.4018/979-8-3693-3007-4.ch002

ABSTRACT

This book chapter discusses the methods and applications of scoping and mapping reviews in medical research, explaining their distinct roles and how they relate to each other. Scoping and mapping reviews are both important exploratory tools designed to address broad research questions by summarizing a wide range of evidence. Scoping reviews are useful for providing an initial overview of the available literature, helping to clarify research questions and identify gaps in knowledge, especially for complex topics. In contrast, mapping reviews are effective for visually summarizing and categorizing existing literature, which helps identify themes and gaps in research. The chapter outlines the step-by-step processes necessary for conducting these reviews, from defining the research question to extracting data and presenting results. It also covers the advantages and challenges associated with each type of review, highlights the importance of involving stakeholders, and discusses the strategic use of different data extraction tools to improve the accuracy and relevance of the results.

INTRODUCTION

Two approaches to assessing large quantities of information in medicine are scoping and mapping reviews. When faced with a project, attempting to learn about the broadness of a research question is the first step in understanding whether to use a mapping review or a scoping review. Scoping reviews mostly help answer open questions with concepts of multiple different topics, whereas mapping reviews help answer more close ended questions where it is required to learn one specific topic.

A scoping review is a type of review that provides a scope of the existing research and can help identify gaps in current literature (Grant & Booth, 2009). Scoping in general is when specific boundaries are designated when completing a project. When defining these specific limits or boundaries there are key boundaries that are placed in order to limit the accessibility and visibility of specific aspects of things to get the job done more efficiently. In the healthcare field researchers apply scoping reviews in order to categorize how significant specific literature is and can be used when gaining medical knowledge (Pham et al., 2014). Scoping is very topic based and essentially leads to learning specific topics only. Indeed, these types of reviews are best designed when attempting to learn specific data about something that is this broad. Thus, by using scoping reviews and making conceptual boundaries of what you are to learn, this will eventually help to bridge the gap in knowledge amongst existing research and your audience.

On the other hand, a mapping review focuses on visually understanding data through a question based format and it pertains to particular concepts. The process of mapping utilizes comprehension and concentration skills, which is usually taken upon the form of note-taking. It connects each fact or idea to one another, creating a graphical representation of lecture content that is quicker to learn for many. This method encourages active engagement, provides instant feedback on understanding, and highlights critical thinking skills (Framingham State University, 2018). Mapping reviews take on a broader approach, which includes a wide range of literature sources and a more inclusive nature. This inclusivity ranges from collecting data from multiple databases, and being time efficient in systematically summarizing key characteristics and cataloging into groups based on publication type, main findings, methodology, etc. In the field of healthcare mapping reviews can be used for numerous different aspects in order to make data management better, improve patient care, and more. This chapter will be focused on the benefits and nuances that both scoping and mapping reviews entail, as well as tips and tricks on how to write them.

EVOLUTION OF SCOPING AND MAPPING REVIEWS

Understanding the development of scoping and mapping reviews is vital in order to understand their current applications and significance.

Early 21st century, research was primarily focused around narrative summaries of existing studies and literature. A structured approach was necessary and researchers realized the lack of such methods. Scoping reviews first emerged in the early 2000s, as the need for a more comprehensive and systematic approach of obtaining evidence arose. One of the original frameworks for conducting a scoping review was proposed by Hilary Arksey and Lisa O'Malley. Arksey and O'Malley established a foundational framework for conducting scoping reviews in their 2005 paper, outlining five key stages. These steps are identifying the research question, identifying the relevant studies, study selection, charting the data, collating, summarizing, and reporting the results, respectively (Arksey & O'Malley, 2005). This was a significant moment that distinguished scoping reviews from regular systematic reviews. The nature of scoping reviews allowed researchers to focus on the broadness rather than the depth of evidence, which popularized scoping reviews.

The progression of scoping reviews over the years can be due to the increase in technological and methodological advancements. Conducting literature searches has become streamlined through digital databases. Search algorithms and tools such as Endnote and Covidence have allowed for efficient data management and analysis. Methodological advancements have structured systematic and scoping reviews with the introduction of PRISMA-ScR extension, which has made the process rigorous and precise (Levac, Colquhoun, & O'Brien, 2010).

Mapping reviews originated in the early 2000s as well, aiming to categorize existing literature into broader topics. Mapping reviews were further developed with the help of Maria J. Grant and Andrew Booth, who devised a framework for mapping reviews. Grant and Booth noticed the lack of categorization of research evidence and took this to their advantage by implementing the evolving trends in their own healthcare work. Grant and Booth were able to conduct high quality reviews by summarizing the most important findings and eliminating unnecessary material (Grant & Booth, 2009, p. 92). Together, they developed an extensive typology that transformed the realm of evidence synthesis. In 2009, Grant and Booth once again collaborated and as a result, the typology became simpler for beginners to understand as well.

The methodologies of mapping reviews became more extensive and detailed over time. Researchers were able to use the typology in order to tailor their approaches to their topics. The original work of Grant and Booth can be used as a reference guide but over time many updated guides were created, referencing their original work (Grant & Booth, 2009). Similar to scoping reviews, conducting mapping reviews became greatly influenced by technological and methodological advancements. As stated before, extensive databases become widely available for research and it has become easier to narrow down sources with enhanced search algorithms.

The future of scoping and mapping reviews remains bright as the usage for them exists in all domains of research. When exploring broad and diverse pools of literature and identifying gaps in knowledge, scoping and mapping reviews will deem necessary. The systematic and methodological precision and rigor allows for reliable and valid results. Scoping and mapping reviews are still utilized to this day due to their adaptable structures to different fields, such as healthcare and education (Levac, Colquhoun, & O'Brien, 2010). Scoping and mapping reviews are valuable tools in today's research practices.

More specifically, scoping reviews have an important function in documenting core concepts that underpin research in a given field and in casting a wide net to elicit some of the most general themes and broad ideas that could be the basis for future research, but which are not yet developed (Peterson et al., 2017). Doing so can alert the scholar to areas of emerging inquiry, conceptual gaps, and potential lines of inquiry (Munn et al., 2022). The critical function of scoping reviews in showing where there are gaps, where much is known, and where almost nothing has been studied is especially valuable when the existing research on the topic is fragmented and rare (Peters et al., 2021). In doing so, scoping reviews inform us about what we know, while also serving as conceptual maps that establish where the body of knowledge on a particular area is thinnest or where it is missing completely (Peters et al., 2015). The wide net cast in a scoping review often elicits disparate parts and the synthesis of information from diverse sources uncovers novel pathways to unexpected connections among concepts and aids in increasing the potential to create interdisciplinary connections between ideas (Verdejo et al., 2021).

However, while scoping reviews aim for breadth, mapping reviews aim for depth in how they organize existing data (Khalil & Tricco, 2022). In contrast to scoping reviews, mapping reviews—sometimes called systematic mapping, or literature mapping reviews—tend to be more targeted in their categorization of findings from existing research (Chen et al., 2023). Although both scoping and mapping reviews can be useful for providing an overview of a research area by summarizing the existing literature, mapping reviews are focused on answering specific questions about the state of the evidence by classifying the literature systematically (Khalil et al., 2024). Mapping reviews typically entail constructing a 'map' summarizing the research landscape, usually in the form of a visual representation, or, occasionally, a dictionary, matrix, or tree (Campbell et al., 2023). However, in all cases, the classification of studies relies on a predefined set of relational criteria that categorize particular studies according to certain key features. For example, studies can be classified by methodological approach, evidence type (randomized controlled trials versus qualitative studies), study population, or the intervention under investigation (Palese et al., 2022). This approach to the literature allows for a typology of the existing literature, such that we can draw conclusions about the concentration of methods, clusters of evidence, and patterns of output, for example (Miake-Lye et al., 2016). For funders and policymakers, having a map of the types of evidence gaps in the literature can inform research priorities, thus coordinating where to invest limited scientific resources to achieve the highest possible impact research output in each discipline of study.

COMPARISON OF THE TYPES OF SYSTEMATIC REVIEWS

There are a multitude of reviews to choose from when deciding upon how to conduct one's research (Grant & Booth, 2009). It is important to consider how the types of reviews fit and serve different research purposes. The following table displays a comparative overview of different types of systematic reviews and their characteristics and expected outputs.

Table 1.

Type of Review	Description	Scope	Output
Systematic Review	Searches, evaluates, and synthesizes research evidence in a systematic manner.	Narrow, focused on a specific question or hypothesis.	Detailed synthesis, meta-analysis, and conclusions based on high-quality evidence.
Critical Review	Aims to show that the writer has done a thorough analysis and critical research on the literature. Extends beyond simple description to incorporate intellectual originality and a level of analysis.	Broad to narrow, dependent on the topic	Critical analysis, identifying strengths, weaknesses, and gaps in the literature.
Literature Review	Published works that offer analysis of recently published or currently available literature.	Broad, covering a wide range of topics	Summary of existing research, identifying key themes, trends, and research gaps.
Mapping Review/ Systematic Mapping	Determine gaps in the research literature and use this information to map out and categorize the current literature in order to propose additional research.	Broad, with emphasis on categorization and visualization of evidence.	Evidence maps, diagrams, and visual representations showing the distribution of research.
Scoping Reviews	An initial evaluation of the possible scope and size of the research literature currently available. Seeks to define the type and volume of research evidence, typically encompassing ongoing studies.	Broad, covering a wide range of studies and evidence types.	Descriptive summary, identifying gaps, and highlighting research areas needing further investigation.
Rapid Review	Evaluation of knowledge regarding a policy or practice issue through the use of systematic review techniques to locate and analyze previous studies.	Narrow to moderate, focused on urgent or time-sensitive questions.	Concise synthesis and recommendations, aimed at providing quick insights.

continued on following page

Table 1. Continued

Type of Review	Description	Scope	Output
Umbrella Review	Relates to the process of gathering evidence from several reviews and organizing it into a single, easily accessible document. Focuses on a broad issue for which there are conflicting options, highlighting evaluations that discuss these options and their outcomes.	Broad, synthesizing results from existing systematic reviews.	Summary of systematic reviews, providing an overarching synthesis and recommendations.
Meta-Analysis	Methodology that provides a more accurate effect of the results by statistically combining the results of quantitative research.	Narrow, focused on studies with comparable data.	Quantitative synthesis with pooled estimates and statistical analysis.
Narrative Review	Provides a comprehensive summary of the literature on a topic, often with a qualitative approach. Often follow a practical approach and are non traditional.	Broad, not restricted to specific study designs.	Comprehensive narrative summary, often highlighting the author's perspective and interpretation.

Note. This table is a culmination of different types of reviews and provides basic insight as to what each review entails. Adapted from "A typology of reviews: an analysis of 14 review types and associated methodologies" by Maria J. Grant & Andrew Booth, 2009, *Health information and libraries journal*, 26(2), 91–108. https://doi.org/10.1111/j.1471-1842.2009.00848.x

Understanding what differentiates each type of review can be quite useful when selecting the most appropriate methodology for a given research question. Each review offers unique insights and results that are tailored to meet specific needs of researchers. Carefully considering the scope and methodology will allow researchers to effectively address their research prompts. This comparative figure serves as a fundamental overview of some of the common types of reviews and aids in making decisions for which review is best suited towards specific research goals.

APPLICATIONS OF SCOPING AND MAPPING REVIEWS: POLICY, PRACTICE, AND FURTHER RESEARCH

Scoping and mapping reviews are valuable tools and contain many applications in the real world. Some of these include policy making, practice, and facilitating future research. These reviews can be helpful in policy making by comprehensively collecting and identifying gaps in literature. By searching a wide database of literature and synthesizing results through scoping and mapping, policymakers will have a better idea of the current state of research. This can further influence decision making based on what information is available and what information is still yet to be discovered. Evidence-informed policymaking (EIPM) is a very important form that requires systematic approaches in order to identify and integrate evidence from various sources (Jabali et al., 2024).

Policymaking in healthcare is a prevalent field that scoping and mapping reviews are used in. Healthcare policy is an essential part of healthcare systems all around the world. Healthcare policy encompasses the choices, objectives, and measures that dictate the way in which healthcare is provided and obtained (Mozafaripour, 2022). For instance, a scoping review of mental health prevention and intervention initiatives for infants and preschoolers at risk for socio-emotional difficulties can help influence healthcare policy (McLuckie et al., 2019). This type of review can provide policymakers with evidence on effective mental health interventions specific to infants and preschoolers. This can further lead to the development of national health policies regarding mental health in infants and toddlers, changing methods at home or in the classroom. Another instance when scoping and mapping reviews impacted policy decision making was HIV/AIDS services in South Africa. Major improvements in HIV/AIDS health service in South Africa have occurred due to a scoping review and the addition of policy dialogue. This included lowering antiretroviral drug prices and improving service delivery models (Yimgang et al., 2021). Researchers and civil society organizations incorporated their research findings in the policymaking process, which made a direct impact on their decision making for HIVS/AIDS services.

Scoping and Mapping reviews help in enhancing practice in clinical, healthcare, and educational settings, to name a few (Maggio et al.,

2021). Mapping reviews can help identify gaps in current fields such as delivering healthcare services. This helps find areas for improvement. A scoping review about parent-targeted postnatal educational interventions in low and middle-income countries mapped out existing interventions (Dol et al., 2019). The aim was to analyze the strengths and limitations of the literature, providing insight as to what works and what does not in practice. This review not only provides a guideline as to what can be improved in practice but it pinpoints critical research areas. The review identifies many gaps pertaining to the educational strategies used for parent-targeted postnatal education. Additionally, the study emphasizes the importance of additional research to identify the best practices and strategy for these interventions. This leads to the idea of how scoping and mapping reviews can facilitate further research. As we know, mapping and scoping reviews can help identify gaps in literature and this has many benefits for prioritizing areas of future research. Furthermore, these reviews can give researchers an idea of where to direct their future research efforts. For example, a mapping review about the impacts of health care service changes implemented due to COVID-19 on children and young people with long-term disability displayed the current research as well as pinpointing many critical research areas (Merrick et al., 2023). The study on childhood disability service adaptations during the COVID-19 pandemic highlighted areas that require further investigation. The effectiveness of support services and other social care adaptations are something that needs to be further researched in children. Additionally, the topic of telehealth in the review suggested that further research would be necessary in order to assess its effectiveness across different conditions and disabilities. Addressing these critical research areas would facilitate further studies and research in the future.

Scoping and Mapping Reviews: Similarities and Differences

Scoping and Mapping reviews are very similar in the way they aid the researcher, they aim to answer big picture research questions. Both types of reviews are exploratory and they map the spectrum of available evidence, in order to gain a comprehensive summary of evidence present pertaining to the question at hand. In turn, both reviews are inclusive and accept literature from diverse studies and methodologies. Mapping and scoping reviews are also less systematic and do not focus on extracting data in a meta-statistical approach. They emphasize more on summarizing and categorizing information to provide an overview of the literature (Campbell et al., 2023). There is also involvement of stakeholders in both types of reviews. Stakeholders are people, groups, or organizations with a stake in the subject being studied or who might be impacted by the review's conclusions. Stakeholder participation is frequently encouraged in both types of reviews in order to improve the impact, relevance, and applicability of the review findings (James, Randall, Haddaway, 2016). Although there are many similarities, there are small details that set both of these reviews apart. Mapping reviews are more broader focused on broad, deductive questions. Their aim is to classify data and characterize patterns through answering targeted questions. They involve a high level of extraction with predefined coding categories, focusing on descriptive information. They are more superficial and use a greater use of visual displays of findings. Scoping reviews attempt to give an overview of the evidence that is currently available on a given subject. Scoping reviews are much more in depth and take a narrower approach on a broader, inductive or deductive question, often tracking underlying concepts. Overall scoping reviews enables a more thorough analysis of the literature, offering a detailed comprehension of ideas and patterns in research (Khalil & Tricco, 2022).

Scoping Review: Benefits and Nuances

Scoping, by definition, is to investigate or assess, and that is exactly what defines the purpose of a scoping review. Scoping reviews are particularly helpful when an ambiguous research question is given due to their exploratory nature (Munn et al., 2018). They are useful because researchers can utilize this type of design for identifying key concepts, terminology, sources, and for [filling] possible gaps in literature. Scoping serves as sort of a preliminary research measure. Scoping review methods can not only be utilized to investigate questions that are open-ended but also act as a starting point for other potential inquiries (Mak et al., 2022). Furthermore, it can help determine the possible criteria by utilizing a variety of sources, incorporating many diverse perspectives. Scoping reviews are intended to map out existing evidence and can be thought of as a precedent to systematic reviews (Gupta et al., 2023). This further helps refine broad research questions and can be fundamental in the initial stages of research. Some of the nuances that can be argued with scoping reviews stems from their benefits. The methodology for some is too heterogeneous, with inclusivity of many diverse literature possibly challenging the validity, quality, and relevance of sources. Although they contain both qualitative and quantitative aspects, a scoping review may be less favored due to absence of risk of bias assessment and limited use of meta-analysis (Bowden & Purper, 2022).

Scoping Review: Methodology Overview

Writing a good scoping review involves a systematic and structured approach. Following a meticulous and thorough process will ensure one to articulate a high-quality scoping review. Scoping reviews start by clearly defining and reviewing objectives and questions that the project or research is attempting to research on. Unlike mapping reviews, scoping reviews have specific boundaries and parameters that data must meet which makes mapping reviews better suited for broader purposes. This results in scoping reviews to be understood quicker as it covers specific concepts. Once clearly defining the research question, an inclusive protocol must be created to identify the review's purpose, and inclusion and exclusion criterias as these criterias are what make this review a scoping review (Sharma & Goyal, 2023). A search strategy with a strongly focused research question and keyword searches is necessary to ensure the search is comprehensive. Next, the systematic literature review must be conducted, across multiple databases. This step is unique from using mapping review as scoping reviews use a synthetic literature search. Once the systematic literature search is completed, all the data found must be screened and only select studies should have data extracted from it to include. This will ensure the data extracted is relevant to the research question. Afterwards, all the data from the extracted group must be organized, interpreted, and summarized to be clear and easily accessible to viewers. A preferred reporting guide can help with these reviews in order to ensure the quality of the review, as all reviews go, it is beneficial to get peer reviews as well (Tricco et al., 2016). The methodology of scoping reviews will be discussed further in detail.

Scoping Review: Step-by-Step Process

The first and foremost step to begin writing a scoping review is to have an objective. It is important to consider what the purpose and aim of the research paper is. From this initial step, stems the research question. The topic is what guides the researcher to conduct research. Other preliminary steps include searching databases to see whether or not scoping review papers of this nature/topic have been published recently. This step will create the foundation for the rest of the scoping review.

The next step would be to develop a protocol, which is a vital component of scoping reviews. A *priori* protocol is an example of a protocol that could be developed, as it defines the framework and search strategies for the scoping review (Peters et al., 2020). This step helps to determine inclusion and exclusion criteria, and predefined components of the review such as the objective, methods, data synthesis, and more, further improving the comprehension of the review. Many reputable sources have templates they provide such as PRISMA's (Preferred Reporting Items for Systematic Reviews and Meta-Analyses) template, JBI Evidence Synthesis's protocol, PROSPERO, and many more. Additionally, the registration for the protocol is necessary to avoid duplication issues and to increase clarity of the review (Straus & Moher, 2010).

One of the major steps of this process is conducting the literature search. As stated before, inclusion and exclusion criteria must be decided in order to set boundaries regarding the study designs, publication date range, relevancy, and type of publications. Searching various databases, such as PubMed, JSTOR, Scopus, Embase, and etc., with specific keywords and Boolean operators (OR or AND) will allow researchers to find sources that are relevant to the topic in order to report on the scope. Extraction of data and literature should be primarily focused on the research topic. The specific sources that are extracted can be used to define, clarify, expand upon, surrounding a concept (Pollock et al., 2023). Other sources can include data that has been collected by other researchers with charts, tables, and graphs, that can support one's paper. Data extracted can also include different study characteristics or other contextual information pertinent to the research. This step does include a lot of reading and searching for details within articles, so developing a strategy will allow researchers to maintain consistency and reliability. Exploring other searching methods such as hand searching and searching grey literature can help to retrieve comparable results (Peters et al., 2020). Ultimately, it is important to rigorously evaluate data from multiple sources to ensure the transparency of the scoping review.

The following step is to organize the citations and screen the evidence. It is crucial to maintain an organization of the extracted citations in order to avoid duplicates and other errors. Screening is the process of determining whether or not an article meets the desired inclusion criteria. Screening through hundreds of articles is a meticulous and lengthy task. Studies show that one of the most efficient methods of screening is based on titles and abstract screening rather than full text examination. This approach has been claimed to be less time consuming and has better screening performance (Teo et al., 2023). This ensures that adequate information is gained to determine whether or not an article should be included. This is a more superficial method and considered the primary level of screening. A more in-depth, secondary level of screening can be done to further include or exclude sources. This will truly make the researcher question "how does this article support my paper?" The criteria remains the same as before but the review and screening process is more rigorous. When working among a team, screening can be split among members to expedite the process. To further assist in the screening process, there are many tools that can be utilized to collaborate and streamline the process of screening. These tools include Rayyan, Covidence, Endnote, Abstrackr, and many more. An efficient screening method further helps to eliminate the risk of bias from researchers when reviewing the articles.

The next step would be to conduct a quality assessment of the findings. The process of searching the data and screening data has been concluded and the next step is to analyze the articles and interpret in what way the articles will answer the research question or fulfill the purpose of the review. Researchers should question the relevance, reliability, validity, and applicability of the included studies (University of North Carolina at Chapel Hill, 2024). One key aspect of a scoping review is that they do not attempt to assess the sources in depth like a systematic review. Scoping reviews are more geared towards mapping and summarizing the findings but the way data is analyzed is ultimately dependent on the author and their needs (Peters et al., 2020). Quality assessment tools and data collection forms are invaluable and help assess the quality of the various studies. Some of the popular quality assessment tools include Cochrane Risk of Bias 2.0, CASP Appraisal checklists, and LEGEND Evidence Evaluation Tools (University of North Carolina at Chapel Hill, 2024). These tools can be applied to a multitude of study designs.

To be more specific, one of the most important tools applied to quality assessment of scientific studies is the Risk of Bias 2.0 (RoB 2.0), which was developed by Cochrane to apply to randomized controlled trials (Higgins et al., 2011). It evaluates bias in the randomization process by examining whether the randomization was adequately performed, if participants were treated according to their randomized allocation using intention-to-treat principles, and whether the outcome measurement aligned with the planned approach (Intention-to-treat analysis, n.d.). Additionally, it assesses the potential for ascertainment bias if a validated and reliable measurement instrument was not used, as well as the selection of the reported results (Spencer & Brassey, 2017). For instance, for the randomization process, RoB 2.0 will evaluate whether there were difficulties with generating the allocation sequence, concealing it from research participants and personnel, or handling the sequence, as each issue could introduce bias by not adequately compensating against randomness (Savović et al., 2014). Likewise, for measurement of the outcome, this domain evaluates the reporting of all outcome data, rather than just a subset, and that the outcome was valid and reliable, because otherwise final analysis could be selective if the outcome was bad in certain groups that were excluded. In total, the RoB 2.0 tool is used to assure the quality of clinical trials in systematic reviews and meta-analyses to determine whether RCTs are sufficiently high-quality and their findings are sound (De Cassai et al., 2023).

Other commonly used tools for critical assessment of bias are the Critical Appraisal Skills Program (CASP) Appraisal Checklists, which are used for a wide variety of different study designs, including qualitative research, cohort studies, case-control studies, and diagnostic test studies (CASP, n.d.). All of the CASP Checklists provide a series of questions designed to be asked of a study, and they are each tailored to the context of the study design under question. To illustrate, the CASP Checklist for qualitative research encompasses questions on whether the aims of the research are clearly articulated, whether the methodology was appropriate, whether the key strengths and limitations of the study are clearly spelled out in the paper, and whether the findings are accessible (Shaheen et al., 2023). Again, these checklists are used to facilitate the critical evaluation of evidence presented at the study level in order to make decisions about care and service delivery based on high-quality research.

LEGEND (Let Evidence Guide Every New Decision) evidence evaluation tools, designed by researchers at the University of North Carolina at Chapel Hill, assess the quality of evidence and the extent to which scientists can rely on the findings of a research study methodologically, or if the findings are generalizable to other populations (Clark et al., 2009). Evidence-evaluation tools are categorized for specific study designs (e.g., LEGEND tools for randomized trials and observational studies as well as systematic reviews). For instance, LEGEND quality assessment for randomized trials evaluates factors such as study design, whether there is a control group, the ability to randomize, sample size, and allocation concealment. This approach also considers the presence of heterogeneity or outliers, consistency, indirect evidence, and the overall grade of the quality of evidence (Clark et al., 2009).

Once the bias has been properly assessed using these objective analyses, the final step of writing the scoping review would be to present the findings. The researcher must decide upon how the data should be presented and how it will effectively communicate to the target audience. Findings can be presented in tables, charts, diagrams, or even in narrative format (Stony Brook University, 2024). It is important to ensure that the tabular data is aligned with the overall objectives and purpose of the study. Descriptions and summaries of the diagrammatic and tabular data should be utilized in conjunction to reflect the objectives of the research (Peters et al., 2020). Grouping the results into major conceptual categories, with concise descriptions of each category published, including research gaps, population, intervention type, duration, goals, methods, and major findings is also recommended. All things considered, this strategy aids in the clear and useful organization and presentation of the scoping review's conclusions (Aromataris & Munn, 2020). Furthermore, guidelines for scoping reviews exist online that can help along the process of writing a scoping review. The PRISMA extension for scoping reviews (PRISMA-ScR) is a 22 item checklist that highlights each section for a scoping review. This can be extremely helpful when writing a scoping review, as it can be checked along every step till completion. The PRISMA website also includes tip sheets for each of the 22 items in the checklist, providing great detail (Tricco et al., 2018). Another great source would be the *JBI Manual for Evidence Synthesis*, which contains extensive information dedicated to systematic and scoping reviews, providing great tips and tricks.

A scoping review manuscript typically follows this order: title, abstract, introduction, methods, results, discussion, funding, appendix, and references. It is important to consider all aspects when constructing a scoping paper and refine the process. Writing a scoping paper is an iterative process based on collaboration and reviews from peers. By following these steps, researchers can effectively write a scoping review that provides a comprehensive overview of existing literature within a specific field or topic area.

Mapping Review: Methodology Overview

Although scoping reviews and mapping reviews show similarities, the methodology of a mapping review is quite different from that of a scoping review. Whereas the scoping review would start off with the purpose of a specific topic, a mapping review starts with clearly defining the purpose of the review and the specific area that is being reviewed. In order to perform a systematic mapping study, a well made plan of the study and search execution should be made to save time and gather simpler inclusion criteria. Next, the scope and purpose of the research or project should be addressed through combing through literature databases and determining boundaries of what information will be included or excluded. Once numerous literature pieces are identified and reviewed, it is important to be clear in which sources are needed based on specific dates, locations, etc. of the source (Croot et al., 2020). Next, data extraction and classification should occur. After the findings are organized it is important to acknowledge the strengths and limitations of existing research in order to evaluate the quality. Furthermore, a synthesis and analysis of these findings is crucial to extract meaningful insights and find specific areas of knowledge gaps to improve on in the future. The methodology of mapping reviews will be discussed further in detail.

Mapping Review: Step-by-Step Process

The underlying principles and strategies used for the step by step process of the scoping review can be applied to the mapping review as well. The tips for literature search and data extraction and analyzing the sources still remain the same with a few differences. Scoping reviews are more broad, primarily aiming to discuss literature pertaining to a specific topic.

Mapping reviews are useful when observing gaps in literature and effective when a specific objective for a more targeted evidence product has not yet been determined. A mapping review is a bit more specific than a scoping review, as the first step would be to gain a clear understanding of the objectives of the mapping project or research (Grant, et al., 2009). Defining what you aim to achieve through the mapping review such as identifying gaps in literature and assessing the accuracy of data. This shifts the focus to a research question [and why a gap exists in the literature], whereas a scoping review is more broad and can address a topic or a specific research question. A helpful aspect is the similar nature of scoping and mapping reviews, that allows users to utilize the PRISMA extension for scoping reviews checklist (PRISMA-Scr). It can be used as a guideline for the mapping review while meeting specific requirements unique to the mapping reviews. A *priori* protocol is greatly recommended before undertaking a mapping review (University of Houston, 2023).

Once the objectives are clearly defined. The scope of the mapping review must be identified through conducting a complete literature search by searching utilizing key words in different databases with consideration to both peer reviewed and gray literature to make sure the most comprehensive primary studies are covered (Croot et al., 2020). This will result in a review that is higher quality. The inclusion and exclusion criteria must still be set in order to eliminate biases and ensure that the sources are up to par. When considering the literature it is important to consider different themes and patterns that appeared among the sources. This can help in the steps down the road to group certain themes or concepts.

The next step is to classify the relevant documents and comprehend the data. All the sources should be reviewed to eliminate inconsistencies and to make sure it is specific to the boundaries of the mapping review. Once the specific sources are categorized and identified it is important to summarize all of the literature compiled into an organized manner, such as chronologically, and it is helpful to add visual aids to do so as it is important that mapping reviews are clear and concise. As stated earlier, once the relevant documents are classified and extracted, the findings must be synthesized and analyzed in order to bring together the main results and conclusions from the selected sources, bridge together gaps of knowledge, and fix discrepancies in the review. It is also key to identify these gaps of knowledge and give a critical reflection of the overall quality of the literature. Proper techniques and strategies should be implemented in order to narrow down the broad pool of literature. This can be done through screening, which helps decide which sources should be excluded and which should be included for the review. This can be streamlined through some of the resources listed before such as Rayyan (Ouzzani et al., 2016).

A tool that is quite essential for mapping reviews is a PRISMA flow diagram. The PRISMA flow diagram is a great tool that will report the records identified, screened, and were included in the end. This is respectively categorized in three steps: identification, screening, and included, with sometimes an additional "eligibility" before included (Rethlefsen & Page, 2022). The PRISMA flow diagram is particularly helpful in showing the flow of information and the number of records from each phase of the literature search and data extraction. This helps maintain transparency and clarity regarding the selection of studies for the mapping review in a graphical aspect. In addition to this, quality assessments for study bias and validity can also be conducted.

The aspect that is more unique to the mapping review is the presentation of the data in the paper. Although the structure of the review remains similar with the title, abstract, introduction, methods, results, discussion, funding, appendix, and references. Mapping reviews are focused on creating visual representations of the findings. In scoping reviews, graphical representations were used in conjunction with text, and the synthesis of information was different as well. For mapping reviews, displaying data is done in a diagrammatic, tabular, flow chart manner (Cooper, 2016). A recently popular visual display that has been used more commonly in mapping reviews are evidence gap maps. Evidence gap maps highlight where research is lacking in literature and it is often organized in a systematic or thematic way in order to make it user friendly and simple to understand (Saran & White, 2018, p. 10). Visual displays utilized in mapping reviews can be quite unique and creative as it is more flexible to represent data and information. Heat maps are another form of visuals used to display volume or strength of evidence which can be useful when identifying clusters and gaps (Ryan et al., 2019, p. 5). Heat maps represent data intensity in a grid format by using color gradients. This effectively visualizes areas of high and low concentration of research. As mentioned before, flow charts are another effective way to display the evidence selection through each stage, and the most commonly used is the PRISMA flow diagram. Flow charts are extremely helpful for outlining the research processes. Another widely used graphic is a bubble chart, which has a similar effect as the heat map. The size of the bubbles can signify quantitative (numerical) data/values or other information based on distance and location along the plot (Morris, 2021). They can also show relationships between variables and display the distribution and quality of evidence across topics. These are only a few visual displays that can be used in mapping reviews. Visual displays overall enhance the information being conveyed and it makes it simpler for the general audience to visualize and comprehend. There is no limitation to the visual representations as long as the information is reliable and comprehensive. The mapping reviews can also draw thematic connections using words as well, aiming to bridge the knowledge gaps by answering the research question, although they are less narrative in nature (University of Houston, 2023). One thing that distinguishes mapping reviews is the fact that they are representing the findings, not directly discussing them. The purpose is to establish connections and links between concepts but to not create a concrete conclusion upon a research question/topic (Cooper, 2016).

Mapping Reviews: Benefits and Nuances

Mapping reviews are similar to scoping reviews in nature. The primary goal is to collate, describe, and categorize current evidence (Campbell et al., 2023). They do not focus on answering a specific question but rather collecting evidence to support the research. Mapping reviews are an excellent way to visualize the literature present pertaining to a research question and can aid in understanding the extent of knowledge. Furthermore, this can help researchers find gaps in the current literature or topic, outlining what research has already been conducted or found and what needs to be done for the future (Cooper, 2016). Similar to scoping reviews, mapping reviews can serve as a precursor to more narrow reviews, such as systematic reviews. They can be utilized to grasp what current information is available in the field and create criteria for a project, before embarking on a more specific investigation. Presentation in mapping reviews can also include visual aids as a method of portraying quantitative data or representing thematic patterns or concepts. Although mapping reviews pose many benefits, there are some drawbacks that may make it infeasible. The initial challenge is creating a plan, especially when the language is used inconsistently in the literature. Linguistic ambiguity might make it more difficult for researchers to index and distribute their work broadly, which can make it harder to reach a wider audience (Campbell et al., 2023). Stakeholders', groups vested in the research, impact can be a reason for conflict when they are involved in mapping reviews. Once again, since mapping reviews are not focused on answering a specific research question, bias may be present while choosing studies and extracting data. The overall dependability of the mapped evidence may be impacted by the inclusion of biased or less rigorous research as a result of mapping reviews' inclusivity. These are some of the nuances that have to be kept in mind when constructing a mapping review.

Tools for Data Extraction

Various programs can be used to accelerate the process of conducting mapping and scoping studies and improve the effectiveness of review procedures in the context of the reviews. Programs such as DistillerSR,

Covidence, and EPPI-Reviewer provide extensive platforms for managing and analyzing scoping and systematic reviews. These solutions offer features including form customization, reviewer cooperation, and faster workflows. Web-based programs like Rayyan are useful for people who prefer open-source or easily accessible options, especially during the screening and data extraction stages (Ouzzani et al., 2016). Additionally, NVivo is a valuable tool that qualitative data analysis and supports various types of sources and offers a range of tools. NVivo is useful for integrating and interpreting data on a larger scale and has quite extensive capabilities to identify patterns and relationships within the data (Sinkovics, 2016). Furthermore, spreadsheet programs like Google Sheets and Microsoft Excel continue to be flexible options for manual data extraction, enabling researchers to design customized templates. Programming languages like Python or R can be used with libraries like Pandas or dplyr for more automated data extraction and analysis, the downside is the learning curve for these coding languages. PRISMA tools have been deemed to be very useful and contain many resources to aid in the process of conducting and writing scoping and mapping reviews. For mapping reviews, Microsoft Excel, Adobe Photoshop, and Canva (a free, online software) are some applications to create clean visual graphics. Technology can help increase productivity for researchers and overall enhance the process of creating systematic reviews.

Tools for Visual Displays

Creating visual displays for scoping and mapping reviews can be done manually but this requires more effort and can be time consuming. Certain tools and programs can allow researchers to make visuals that effectively represent the data and evidence in an organized manner. These tools can help generate diagrams, charts, and other images that present their research findings constructively.

A very common tool used for generating charts and diagrams is Microsoft Excel, which is a software free of charge to many researchers (Godino, 2023). Microsoft Excel is a widely known application and it is used to create charts, tables, and graphs. It is quite intuitive and requires simple skills in order to manipulate data making it quite user friendly.

VOSviewer is a software tool that generates graphics using network (or text) data to employ their Visualization of Similarities (VOS) mapping technique (Sinkovics, 2016). VOSviewer is a great tool, particularly for creating evidence gap maps due to its broad support of sources. VOSviewer offers extensive mapping functions and creates high quality visuals seamlessly, making it versatile for different types of reviews (Singapore Management University, n.d.).

NVivo, as mentioned earlier, is a software that is primarily used for data analysis, but it also contains options for generating visuals. NVivo is useful for creating complex node structures and visualizing data through its model-building function (Sinkovics, 2016). This allows for NVivo to generate thematic maps and conceptual maps that help in synthesizing qualitative data.

Some of the programming languages discussed earlier such as Python and R are not only useful for data analysis but they can also be useful for visualization. Applications within the coding languages are what allows for the creation of customizable plots and graphs. Python uses libraries such as Matplotlib and Seaborn and R uses packages like ggplot2 (Singh & Rawat, 2023). Both languages are great for handling large sets of data and can help create complex visualizations for scoping and mapping reviews.

For creating more interactive visuals Tableau and Flourish can be utilized. The greatest benefit to using programs such as Tableau and Flourish is their ability to create highly detailed and interactive visuals. Using interactive platforms can be a great way to present multiple variables in a dynamic manner making it suitable for more detailed visual analysis (South & Rodgers, 2023). Additionally, using interactive graphics can allow researchers to present their data on a different scale and dimension. Which can enhance their data and allow users to be more engaged.

In addition to these applications, there are more niche softwares that can be used to create mapping diagrams. Bibexcel is an open source software that uses bibliometric data and can construct co-occurrence matrices, but the downside is that it requires an external program to provide visualization (Persson, Danell, & Schneider, 2009). CiteSpacell is a program that is excellent in detecting visual patterns and trends within scientific literature (Chen, 2006). IN-SPIRE is a software that uses a landscape metaphor to help users understand relationships among documents. Additionally, IN-SPIRE can provide tools for facet analysis, grouping, and slicing (Sangam & Mogali, 2012).

Choosing the right software for creating visual displays is reliant upon the researcher's ability to learn the software and the goal they are aiming to achieve by creating a certain graphic. The type of visual can greatly impact the effectiveness and clarity of data being presented in a review. Therefore, it is important to consider the appropriate software that can streamline the process and enhance the quality of the visuals to better support the needs of the review. By leveraging these tools effectively, researchers can produce clear and informative visuals that enhance the understanding of their findings.

CONCLUSION/OVERVIEW

When done correctly, scoping and mapping reviews are invaluable tools that give researchers a thorough evaluation of a subject. These two reviews, including the subtleties that set them distinct and the commonalities between them. Along with learning about subtleties, readers will also learn how to combat them when drafting a scoping and mapping review. They will also pick up additional research techniques and data extraction tools to help them write their own scoping or mapping reviews. Scoping and mapping reviews provide a more systematic method to comprehending existing knowledge and assistance in future research attempts as researchers navigate the vast landscape of literature.

REFERENCES

Arksey, H., & O'Malley, L. (2005). Scoping studies: Towards a methodological framework. *International Journal of Social Research Methodology*, *8*(1), 19–32. DOI: 10.1080/1364557032000119616

Aromataris, E., & Munn, Z. (Eds.). (2020). *JBI manual for evidence synthesis*. JBI., DOI: 10.46658/JBIMES-24-01

Bowden, V. R., & Purper, C. (2022). Types of reviews—Part 3: Literature review, integrative review, scoping review. *Pediatric Nursing*, *48*(2), 97–101. DOI: 10.62116/PNJ.2022.48.2.97

Campbell, F., Tricco, A. C., Munn, Z., Pollock, D., Saran, A., Sutton, A., White, H., & Khalil, H. (2023). Mapping reviews, scoping reviews, and evidence and gap maps (EGMs): The same but different—The "big picture" review family. *Systematic Reviews*, *12*(1), 45. Advance online publication. DOI: 10.1186/s13643-023-02178-5 PMID: 36918977

Chen, C. (2006). CiteSpace II: Detecting and visualizing emerging trends and transient patterns in scientific literature. *Journal of the American Society for Information Science and Technology*, *57*(3), 359–377. DOI: 10.1002/asi.20317

Chen, Y. F., Takwoingi, Y., & Grove, A. (2023). Mapping review as a subtype of scoping review: Differentiating between mapping evidence and presentation of evidence as maps. *Journal of Clinical Epidemiology*, *157*, 110–111. DOI: 10.1016/j.jclinepi.2023.02.017 PMID: 36801486

Clark, E., Burkett, K., & Stanko-Lopp, D. (2009). Let evidence guide every new decision (LEGEND): An evidence evaluation system for point-of-care clinicians and guideline development teams. *Journal of Evaluation in Clinical Practice*, *15*(6), 1054–1060. DOI: 10.1111/j.1365-2753.2009.01314.x PMID: 20367705

Cooper, I. D. (2016). What is a "mapping study"? *Journal of the Medical Library Association: JMLA*, *104*(1), 76–78. DOI: 10.3163/1536-5050.104.1.013 PMID: 26807058

Critical Appraisal Skills Programme. (n.d.). *CASP checklists*. https://casp-uk.net/casp-tools-checklists/

De Cassai, A., Boscolo, A., Zarantonello, F., Pettenuzzo, T., Sella, N., Geraldini, F., Munari, M., & Navalesi, P. (2023). Enhancing study quality assessment: An in-depth review of risk of bias tools for meta-analysis. *Journal of Anesthesia, Analgesia and Critical Care*, *3*(1), 44. Advance online publication. DOI: 10.1186/s44158-023-00129-z PMID: 37932825

Dol, J., Campbell-Yeo, M., Tomblin Murphy, G., Aston, M., McMillan, D., Gahagan, J., & Richardson, B. (2019). Parent-targeted postnatal educational interventions in low- and middle-income countries: A scoping review and critical analysis. *International Journal of Nursing Studies*, *94*, 60–73. DOI: 10.1016/j.ijnurstu.2019.03.011 PMID: 30933873

Godino, L. (2023). How to structure Microsoft Excel documents for systematic reviews. *Nurse Researcher*, *31*(1), 40–46. DOI: 10.7748/nr.2023.e1866 PMID: 36856031

Grant, M. J., & Booth, A. (2009). A typology of reviews: An analysis of 14 review types and associated methodologies. *Health Information and Libraries Journal*, *26*(2), 91–108. DOI: 10.1111/j.1471-1842.2009.00848.x PMID: 19490148

Higgins, J. P. T., Altman, D. G., Gøtzsche, P. C., Juni, P., Moher, D., Oxman, A. D., Savovic, J., Schulz, K. F., Weeks, L., & Sterne, J. A. C. (2011). The Cochrane Collaboration's tool for assessing risk of bias in randomised trials. *BMJ (Clinical Research Ed.)*, *343*(oct18 2), d5928. DOI: 10.1136/bmj.d5928 PMID: 22008217

Jabali, S. H., Yazdani, S., Pourasghari, H., & Maleki, M. (2024). From bench to policy: A critical analysis of models for evidence-informed policymaking in healthcare. *Frontiers in Public Health*, *12*, 1264315. DOI: 10.3389/fpubh.2024.1264315 PMID: 38596514

James, K. L., Randall, N. P., & Haddaway, N. R. (2016). A methodology for systematic mapping in environmental sciences. *Environmental Evidence*, *5*(1), 7. Advance online publication. DOI: 10.1186/s13750-016-0059-6

Khalil, H., Campbell, F., Danial, K., Pollock, D., Munn, Z., Welsh, V., Saran, A., Hoppe, D., & Tricco, A. C. (2024). Advancing the methodology of mapping reviews: A scoping review. *Research Synthesis Methods*, *15*(3), 384–397. DOI: 10.1002/jrsm.1694 PMID: 38169156

Khalil, H., & Tricco, A. C. (2022). Differentiating between mapping reviews and scoping reviews in the evidence synthesis ecosystem. *Journal of Clinical Epidemiology*, *149*, 175–182. DOI: 10.1016/j.jclinepi.2022.05.012 PMID: 35636593

Levac, D., Colquhoun, H., & O'Brien, K. K. (2010). Scoping studies: Advancing the methodology. *Implementation Science : IS*, *5*(1), 69. Advance online publication. DOI: 10.1186/1748-5908-5-69 PMID: 20854677

Maggio, L. A., Larsen, K., Thomas, A., Costello, J. A., & Artino, A. R.Jr. (2021). Scoping reviews in medical education: A scoping review. *Medical Education*, *55*(6), 689–700. DOI: 10.1111/medu.14431 PMID: 33300124

McLuckie, A., Landers, A. L., Curran, J. A., Cann, R., Carrese, D. H., Nolan, A., Corrigan, K., & Carrey, N. J. (2019). A scoping review of mental health prevention and intervention initiatives for infants and preschoolers at risk for socio-emotional difficulties. *Systematic Reviews*, *8*(1), 183. Advance online publication. DOI: 10.1186/s13643-019-1043-3 PMID: 31337424

Merrick, H., Driver, H., Main, C., Kenny, R. P. W., Richmond, C., Allard, A., Bola, K., Morris, C., Parr, J. R., Pearson, F., & Pennington, L. (2023). Impacts of health care service changes implemented due to COVID-19 on children and young people with long-term disability: A mapping review. *Developmental Medicine and Child Neurology*, *65*(7), 885–899. DOI: 10.1111/dmcn.15503 PMID: 36649197

Miake-Lye, I. M., Hempel, S., Shanman, R., & Shekelle, P. G. (2016). What is an evidence map? *Systematic Reviews*, *5*(1), 28. Advance online publication. DOI: 10.1186/s13643-016-0204-x PMID: 26864942

Munn, Z., Peters, M. D. J., Stern, C., Tufanaru, C., McArthur, A., & Aromataris, E. (2018). Systematic review or scoping review? *BMC Medical Research Methodology*, *18*(1), 143. DOI: 10.1186/s12874-018-0611-x PMID: 30453902

Munn, Z., Pollock, D., Khalil, H., Alexander, L., Mclnerney, P., Godfrey, C. M., Peters, M., & Tricco, A. C. (2022). What are scoping reviews? *JBI Evidence Synthesis*, *20*(4), 950–952. DOI: 10.11124/JBIES-21-00483 PMID: 35249995

Ouzzani, M., Hammady, H., Fedorowicz, Z., & Elmagarmid, A. (2016). Rayyan—A web and mobile app for systematic reviews. *Systematic Reviews*, *5*(1), 210. Advance online publication. DOI: 10.1186/s13643-016-0384-4 PMID: 27919275

Palese, A., Mansutti, I., Visintini, E., Caruzzo, D., Moreale, R., Longhini, J., & Danielis, M. (2022). Framing the time while designing and conducting reviews: A focused mapping review and synthesis. *Journal of Clinical Nursing*, *31*(23–24), 3523–3534. DOI: 10.1111/jocn.16180 PMID: 34936157

Persson, O., Danell, R., & Schneider, J. W. (2009). How to use Bibexcel for various types of bibliometric analysis. Celebrating scholarly communication studies: A Festschrift for Olle Persson at his 60th Birthday, 5(2009), 9-24.

Peters, M. D. J., Godfrey, C., & Mclnerney, P.. (2020). Scoping reviews. In Aromataris, E., & Munn, Z. (Eds.), *JBI manual for evidence synthesis*. JBI., DOI: 10.46658/JBIMES-20-12

Peters, M. D. J., Godfrey, C. M., Khalil, H., Mclnerney, P., Parker, D., & Soares, C. B. (2015). Guidance for conducting systematic scoping reviews. *International Journal of Evidence-Based Healthcare*, *13*(3), 141–146. DOI: 10.1097/XEB.0000000000000050 PMID: 26134548

Peters, M. D. J., Marnie, C., Colquhoun, H., Garritty, C. M., Hempel, S., Horsley, T., Langlois, E. V., Lillie, E., O'Brien, K. K., Tunçalp, ., Wilson, M. G., Zarin, W., & Tricco, A. C. (2021). Scoping reviews: Reinforcing and advancing the methodology and application. *Systematic Reviews*, *10*(1), 263. Advance online publication. DOI: 10.1186/s13643-021-01821-3 PMID: 34625095

Peterson, J., Pearce, P. F., Ferguson, L. A., & Langford, C. A. (2017). Understanding scoping reviews. *Journal of the American Association of Nurse Practitioners*, *29*(1), 12–16. DOI: 10.1002/2327-6924.12380 PMID: 27245885

Pham, M. T., Rajić, A., Greig, J. D., Sargeant, J. M., Papadopoulos, A., & McEwen, S. A. (2014). A scoping review of scoping reviews. *Research Synthesis Methods*, *5*(4), 371–385. DOI: 10.1002/jrsm.1123 PMID: 26052958

Pollock, D., Peters, M. D. J., Khalil, H., McInerney, P., Alexander, L., Tricco, A. C., Evans, C., de Moraes, É. B., Godfrey, C. M., Pieper, D., Saran, A., Stern, C., & Munn, Z. (2023). Recommendations for the extraction, analysis, and presentation of results in scoping reviews. *JBI Evidence Synthesis*, *21*(3), 520–532. DOI: 10.11124/JBIES-22-00123 PMID: 36081365

Rethlefsen, M. L., & Page, M. J. (2022). PRISMA 2020 and PRISMA-S. *Journal of the Medical Library Association: JMLA*, *110*(2), 253–257. DOI: 10.5195/jmla.2022.1449 PMID: 35440907

Saran, A., & White, H. (2018). Evidence and gap maps: A comparison of different approaches. *Campbell Systematic Reviews*, *14*(1), 1.

Spencer, E. A., & Brassey, J. (2017). Ascertainment bias. In *Catalogue of bias*. https://catalogofbias.org/biases/ascertainment-bias/

Straus, S., & Moher, D. (2010). Registering systematic reviews. *Canadian Medical Association Journal*, *182*(1), 13–14. DOI: 10.1503/cmaj.081849 PMID: 19620270

Teo, L., Van Elswyk, M. E., Lau, C. S., & Shanahan, C. J. (2023). Title-plus-abstract versus title-only screening. *Systematic Reviews*, *12*, 211. Advance online publication. DOI: 10.1186/s13643-023-02374-3 PMID: 37957691

Tricco, A. C., Lillie, E., Zarin, W., O'Brien, K., Colquhoun, H., Kastner, M., Levac, D., Ng, C., Sharpe, J. P., Wilson, K., Kenny, M., Warren, R., Wilson, C., Stelfox, H. T., & Straus, S. E. (2016). A scoping review on the conduct and reporting of scoping reviews. *BMC Medical Research Methodology*, *16*(1), 15. Advance online publication. DOI: 10.1186/s12874-016-0116-4 PMID: 26857112

Tricco, A. C., Lillie, E., Zarin, W., O'Brien, K. K., Colquhoun, H., Levac, D., Moher, D., Peters, M. D. J., Horsley, T., Weeks, L., Hempel, S., Akl, E. A., Chang, C., McGowan, J., Stewart, L., Hartling, L., Aldcroft, A., Wilson, M. G., Garritty, C., & Straus, S. E. (2018). PRISMA extension for scoping reviews (PRISMA-ScR). *Annals of Internal Medicine*, *169*(7), 467–473. DOI: 10.7326/M18-0850 PMID: 30178033

Verdejo, C., Tapia-Benavente, L., Schuller-Martínez, B., Vergara-Merino, L., Vargas-Peirano, M., & Silva-Dreyer, A. M. (2021). What you need to know about scoping reviews. *Medwave*, *21*(2), e8144. DOI: 10.5867/medwave.2021.02.8144 PMID: 33914717

Yimgang, D., Danhoundo, G., Kusi-Appiah, E., Sunder, V., Campbell, S., & Yaya, S. (2021). A scoping review of researchers' involvement in health policy dialogue in Africa. *Systematic Reviews*, *10*(1), 190. Advance online publication. DOI: 10.1186/s13643-021-01745-y PMID: 34174957

Chapter 3
Systematic Reviews of Diagnostic Studies

Mohammad-Salar Hosseini
https://orcid.org/0000-0003-2765-5018
Tabriz University of Medical Sciences, Iran

Mohammad Amin Akbarzadeh
https://orcid.org/0000-0002-9589-104X
Tabriz University of Medical Sciences, Iran

ABSTRACT

Systematic reviews of diagnostic studies are essential tools for synthesizing evidence on the performance of diagnostic tests, providing clinicians and policymakers with valuable insights into test accuracy and reliability. Focusing on the accuracy of the diagnostic tests, these systematic reviews are commonly called the systematic reviews of diagnostic test accuracy (DTA). This chapter provides a detailed overview of these reviews, highlighting their specific methodological characteristics, types, and requirements and providing a thorough practical guide to conducting these studies. The key considerations for conducting a systematic review of diagnostic studies, including the formulation of the study question, protocol development, search strategy development, systematic search, study selection, data extraction, bias in DTA studies, critical appraisal, tools for risk of bias assessment, synthesis and meta-analysis, clinical utility, and implications for patient-oriented research are covered in this chapter.

DOI: 10.4018/979-8-3693-3007-4.ch003

INTRODUCTION

Systematic reviews (SRs) are the pillars of evidence-based medicine, providing a comprehensive and unbiased synthesis of existing research. Among the various types of systematic reviews, diagnostic study reviews hold a unique position. Focusing on the accuracy of the diagnostic tests, these systematic reviews are commonly called the Diagnostic Test Accuracy Systematic Reviews (DTA-SRs). Generally, DTA-SRs are methodologically rigorous investigations that proceed to systematically identify, appraise, and synthesize evidence regarding the performance of novel diagnostic tests in detecting specific diseases or conditions. These reviews also provide insights into the reliability and clinical utility of diagnostic tests. Unlike conventional systematic reviews, which primarily focus on interventions or treatments, DTA-SRs deal with the sensitivity, specificity, likelihood ratios, and other performance metrics of diagnostic tests.

Accurate and timely diagnosis is the foundation of effective patient management in clinical practice. Diagnostic tests are essential tools in disease detection, risk stratification, and selection of treatment strategies. Considering the advantages of data pooling, DTA-SRs provide comprehensive insights into novel diagnostic tests, making it feasible to compare available and novel tests against the current diagnostic gold standards. Moreover, these reviews serve as a foundation for developing clinical practice guidelines, ensuring standardized approaches to diagnostic testing across healthcare settings. Considering the escalating healthcare costs and finite resources, DTA-SRs also provide policymakers with invaluable evidence regarding the comparative effectiveness and cost-effectiveness of diagnostic tests, enabling informed allocation of resources and optimization of healthcare delivery.

This chapter aims to provide an inclusive overview of the DTA-SRs and a practical approach to conducting them.

DEFINITION AND TYPES

Systematic reviews of diagnostic test accuracy (DTA) studies are defined as secondary studies following an evidence-based systematic approach to provide a summarized performance and pooled analysis of diagnostic tests while evaluating all available studies and addressing variation in findings between studies (Leeflang, 2014; Takwoingi & Quinn, 2018)

DTA studies either compare two available diagnostic tests or a diagnostic test of interest, called an "index test", with a currently used one, called a "reference test", in diagnosing a specific disease or condition among a particular population. If the best available diagnostic test for accurately identifying the presence or absence of the condition of interest has already been indicated through previous studies or practice guidelines, the reference test would be referred to as the "gold standard" or "reference standard" diagnostic test. In this case, the sensitivity and specificity of the new tests will be calibrated regarding the diagnostic gold standard.

Following the definition, two main designs for DTA studies could be identified, which are applicable to DTA-SRs (Leeflang, 2014):

1. Single-gate studies (cohort type accuracy studies): Using a single set of inclusion criteria, a set of population, potentially with the disease or condition of interest, undergo both index and reference tests.
2. Two-gate studies (case-control type accuracy studies): Using different sets of criteria, the test results from one set of population (cases) are compared to those of another group (controls).

Although two-gate studies provide significant potential for more studies, they have been generally criticized for their inflated accuracy estimates (Rutjes et al., 2005). Therefore, researchers and clinicians should consider the possible spectrum effects in two-gate study designs. Although pooling both types of DTA studies together for meta-analysis is generally not prohibited, the type of DTA could be filtered through the inclusion/exclusion criteria at the beginning of DTA-SR studies, or a subgroup pooling and analysis based on the type of DTA could be performed further in the study to ensure the reliability of pooled results (Olsen et al., 2019).

KEY DIFFERENCES OF DTA AND NON-DTA SYSTEMATIC REVIEWS

While DTA-SRs share foundational principles with conventional SRs, several key distinctions are evident due to the unique nature of diagnostic test studies. Unlike therapeutic interventions, diagnostic tests use a variety of modalities, ranging from laboratory assays to imaging techniques, each with its own inherent complexities and sources of bias. Another notable difference is the selection and synthesis of primary studies. In DTA-SRs, the focus extends beyond randomized controlled trials (RCTs) to include a broader spectrum of study designs, including observational studies. This inclusive approach accompanies diverse methodologies, necessitating a comprehensive appraisal of test performance across varied clinical settings and patient populations. Moreover, DTA-SRs often face unique methodological challenges inherent to diagnostic tests' evaluation, such as spectrum, verification, and incorporation biases, which require specific statistical and presentation techniques, including receiver operating characteristic (ROC) curve analysis, hierarchical modeling, and bivariate meta-analysis. Apart from the methodological complexity and heterogeneity in test performances, translating diagnostic study findings into clinical practice is another key difference of DTA-SRs with non-DTA-SRs, which involves substantial expertise in research methodology and clinical aspects (Behringer & Freund, 2021).

OUTLINE AND OVERVIEW OF STEPS

DTA-SRs follow a structured framework with several key steps. First, an expert team of researchers should be formed, with an overall acceptable expertise in biostatistics, research methodology, evidence-based medicine, and clinical medicine. Like any other type of SR, study design and protocol development (and registration) are essential initial steps, with each member's role clearly defined. Formulating a well-defined research question lays the foundation, guiding the subsequent stages of the review. Next, a comprehensive and systematic literature search should be conducted across various relevant databases to include all relevant primary studies. Data extraction and quality assessment begin following the screening of search results based on pre-defined inclusion/exclusion criteria. The synthesis of findings – with or without meta-analysis – and the final interpretation of the results lead to the formulation of conclusions.

The following sections discuss each step in practical details:

STEP ONE: STUDY DESIGN AND PROTOCOL DEVELOPMENT

Similar to other SRs, conducting a DTA-SR starts with developing the study design and revolves around the defined study question. The initial idea behind the study question could be triggered by facing a clinical challenge, incompetence of practical but defective available diagnostic criteria, or indicated research gaps. At the outset of a DTA-SR, the formulation of a precise and focused research question serves as the cornerstone of inquiry. A well-defined research question delineates the scope and objectives of the review, guiding subsequent stages of study selection, data extraction, and synthesis. The research question in a DTA-SR typically centers on assessing the accuracy and reliability of diagnostic tests in detecting a specific condition or disease within a defined population.

The Preferred Reporting Items for Systematic Reviews and Meta-analyses

The Preferred Reporting Items for Systematic Reviews and Meta-Analyses (PRISMA) are a minimum set of items for reporting in systematic reviews and meta-analyses, with the latest update released in 2020 (Page et al., 2021). PRISMA helps authors improve the reporting of SRs, and in order to be practical for all types of SRs, several extensions for PRISMA have been introduced over time. The Preferred Reporting Items for a Systematic Review and Meta-Analysis of Diagnostic Test Accuracy Studies (PRISMA-DTA) is an extension of PRISMA, reflecting the specific requirements for pooled reviews of DTA studies (McInnes et al., 2018). Although it principally focuses on reporting DTA-SRs, it also provides valuable general insights for conducting these studies, and should be considered and consulted from the initial stages of the study design (Salameh et al., 2020). PRISMA-DTA will be discussed further in this chapter, but as an initial insight into the guideline, the study protocol should be developed and registered prospectively, with the study objective clearly stated.

Study Question

The study question, also called the research question, is the fundament of the study. A good study question helps identify the review's key components and adapt a suitable approach to the study. Also, a well-structured study question ensures that the knowledge gaps are studied and well-defined prior to the study.

The research questions should be feasible, interesting, novel, ethical, and relevant (also known as FINER criteria) (Willis, 2023). The specificity, clarity, originality, and measurability of the study questions are among other conditions that study questions should meet. Considering the discussed characteristics of DTA studies, the classical frameworks for formulating study questions (PICO) might not be practical for DTA-SRs. Several frameworks – derivates from the classical PICO – have been previously proposed for DTA-SRs, the most famous of them being the *PIRD* framework, which is consisted of the following components (Hosseini et al., 2024):

Population

The population points to the particular target disease or condition, and/or the group undergoing the diagnostic measures should be defined in as much detail as possible.

Index Test

Index test indicates the diagnostic test or the diagnostic measure that is evaluated through primary studies.

Reference Test

The reference test usually refers to the available gold standard diagnostic tool or the best method available for detecting patients with the particular target disease or condition.

Diagnosis of Interest

Diagnosing the particular target disease/condition with the highest sensitivity and specificity is the ultimate goal of DTA studies and should also be incorporated into the primary study question.

Preliminary Literature Review

A brief literature review is essential prior to study design, ensuring the knowledge gaps and the originality and non-duplication of the review. To ensure non-duplication and to decrease the risk of reporting bias, registration of the review protocols is highly recommended (van der Braak et al., 2023).

Protocol Components and Requirements

DTA-SRs are important parts of evidence-based decision-making, and their suboptimal reporting compromises the validity, reliability, and generalizability, affecting their clinical applicability and value. As a part of general measures towards research integrity and transparency, the research teams are encouraged to prospectively register their study protocols. The initial protocols are encouraged to be as detailed as possible. The details vary based on the requirements of each registry and the authors' preferences; however, a research protocol should at least contain details regarding the following aspects of the review:

Team Members

A complete list of team members should be indicated in the original study protocol. Preferably, contact information for at least one person should be mentioned.

Eligibility Criteria and Tests

The eligibility criteria should be clearly defined in terms of inclusion and exclusion criteria for primary study characteristics (such as participants, settings, and tests) and report characteristics (such as publication year and language), provided along with the rationale.

Outcomes and Outcome Measures

Outcomes, usually the previously discussed diagnosis of interest, and their measure, generally consisting of sensitivity, specificity, and positive and negative predictive values, should be indicated as clearly as possible. In case of multiple definitions, the target disease or condition should be clearly defined beforehand.

Quality Assessment

Critical appraisal and quality assessment of DTA-SRs will be discussed in detail in upcoming sections; however, at this point, remember that DTA studies are quite bias-prone, and a clear strategy of critical appraisal and the tools that will be used should be indicated in the protocol.

Data Synthesis

The protocol should highlight the team's plan for data synthesis, generally how the feasibility of meta-analysis is determined, and how data pooling will be performed. The authors should prospectively determine how the thresholds of test positivity will be selected and how the indeterminate test results will be handled. If applicable, subgroups are worth mentioning in the protocol.

Source of Funding

If applicable, the details regarding the funders and sponsors of the study should be declared transparently.

Protocol Registration

Although there has been no obligation on which protocol registry should be used, PRISMA guidelines generally recommend the well-known International Prospective Register of Systematic Reviews (PROSPERO), introduced by the Center for Reviews and Dissemination of York University, funded by the National Institute of Health Research, UK (*About PROSPERO*; Bernardo, 2017). Other registries, such as the International Platform of Registered Systematic Review and Meta-analysis Protocols (INPLASY) (*Learn more about INPLASY*), the Research Registry (*About the Research Registry*), the Open Science Framework (OSF) (*The Open Science Framework*), and protocols.io (*About us - protocols.io*) have also been used by the authors. The institutional repository might also be used to archive the initial protocol only if it supports a public access policy and non-modification of the initially stored protocol is ensured. However, considering the availability of standard protocol registration formats, the complete archive of previously registered reviews with the ability to search, and rapid response and registration, the free-of-cost and publicly-available PROSPERO is of significant popularity (Pieper & Rombey, 2022).

STEP TWO: SYSTEMATIC SEARCH

Like any other type of SR, the systematic literature search of DTA-SRs aims to provide a list of relevant studies as extensive and complete as possible (Liu et al., 2013). This systematic literature search usually includes published primary journal articles, but could also include unpublished literature, conference proceedings, dissertations, and theses, depending on the study question (Deeks et al., 2023). This step could be broken into small steps:

Database Selection

Selecting multiple databases for systematic search improves the coverage and search sensitivity/recall (Ewald et al., 2022). Although guidelines do not recommend a mandatory or definite minimum number of databases for systematic search, the general expert opinion revolves around including at least three independent databases with a wide range of coverage (Mathew, 2024). MEDLINE, Embase, Ovid, Scopus, Web of Science (WoS – previously known as Web of Knowledge), China National Knowledge Infrastructure (CNKI), CINAHL, PsycINFO, ProQuest, and the Cochrane Library are among the most commonly searched databases, each with its advantages and disadvantages. For instance, MEDLINE, from the National Library of Medicine (NLM), has a broad index of journals and citations and is freely accessible through PubMed. Embase and Ovid also have adjunctive access to MEDLINE, but they are only available on subscription plans. The same restrictions come with Scopus and WoS, which need personal/institutional subscriptions for access but provide an inclusive set of results for researchers. Apart from general databases with broad index, some narrow-index databases, such as regional databases (like CNKI) or subject-specific databases (like PsycINFO), could be included among the searched databases with specific objectives, populations, or diagnoses of interest. In summary, a list of potentially more than two databases tailored to the study's objectives and authors' expertise should be selected for a DTA-SR to ensure acceptable recall and inclusivity.

Search Strategy Development

Search Strategy Development has been extensively discussed in other chapters. However, some key points for DTA-SRs are discussed here. DTA studies usually have complex search strategies due to the lack of proper subject headings and related database search filters among study types (Deeks et al., 2023). Also, DTA studies usually lack proper keywords regarding the measures of accuracy used, leading to ineffective indexing and text word searching. To ensure retrieving a complete list of relevant records, including free keywords related to the MeSH terms are also recommended, as their combination in the search strategy has previously demonstrated the highest efficacy in retrieving maximum relevant results (DeMars & Perruso, 2022). The *PIRD* framework could be used to develop the search strategy, maximizing the results by properly using the Boolean operators. For instance, the following Table 1 presents the complete search strategy via OvidSP for a previously published DTA-SR on ultrasound in the diagnosis of sarcopenia (Fu et al., 2023):

Table 1. The complete search strategy via OvidSP for a previously published study on ultrasound in the diagnosis of sarcopenia (Fu et a., 2023)

Search Queries	
#1	sarcopenia/ or muscular atrophy/
#2	(sarcopenia or "muscle atroph*" or "muscular atroph*" or "muscle mass*" or "muscle size*" or "muscle diameter*" or "muscle volume*" or "muscle thickness*" or "muscle wasting").ti,ab.
#3	Ultrasonics/
#4	("Ultrasonic" or "ultrasonography" or "ultrasound" or "sonography" or "ultraso* imaging" or "echography" or "elastography" or "sonoelastography" or "echo intensity").ti,ab.
#5	1 or 2
#6	3 or 4
#7	5 and 6

There are two general approaches to developing search strategies: Broad search strategy (Sensitivity-maximizing approach), and narrow search strategy (specificity-/precision-maximizing approach). The broad search strategy leads to more results, ensuring no relevant study has been missed potentially, whereas the latter leads to fewer results and a rapid screening phase, with the potential risk of missing some relevant studies. Generally, a sensitivity-maximizing approach has been encouraged over the narrow search strategy (Glanville et al., 2020). However, if a high number of records have been captured, the specificity-/precision-maximizing approach could be replaced. Also, remember that if an extremely high number of relevant results have been retrieved, the subject might have already been studied sufficiently and potentially settled; therefore, the need for a new SR in the field might be questionable.

Following PRISMA-DTA, the full search strategy for *all* databases (and other sources) should be reported, along with any limitations in language, time, and database filters, to ensure the results' replicability.

Search Strategy Registration

Registering the search strategy is not a mandatory step. However, since the complete and ultimate search strategy is usually not available at the time of study protocol registration, some researchers have suggested registering the search strategy right after its development, to ensure the lack of bias and non-modification principles considering the vital role of search strategy in integrity of the review. Currently *searchRxiv*, part of the Center for Agriculture and Bioscience International (CABI) provides the search strategy registration services (*About: searchRxiv*). Also, the authors could include search strategies as amendments to their registered original protocols.

Additional Sources

Additional sources are generally recommended to guarantee including all relevant studies and records. The most common additional sources are as follows: 1) References of the references: The references of the studies and records from formal search results could provide a complete list of relevant studies, which is especially useful for screening records that might have been missed due to confusing titles. 2) Search engines: Scientific search engines, such as Google Scholar, have never been included among formal databases since they are not "systematically" searchable and show no consistency in results, leading to non-replicable results. However, they are still a generous source of relevant studies and could be used to improve the inclusiveness of the results. 3) Handsearching: Especially for the fields with recognized flagship journals, handsearching every issue of those journals could help the authors find more studies relevant to the aim of the review. 4) Grey literature databases: Informal publications and documents, such as conference abstracts, policy briefs, and registry reports, are among the grey literature, which could be used to develop a rich data pool. 5) Preprint servers: There is currently no consensus regarding the inclusion of data from preprints for data pooling. However, the general opinion values the integrity of research, using more reliable studies.

STEP THREE: STUDY SELECTION AND DATA EXTRACTION

DTA-SRs require careful study selection and extensive data extraction to ensure no relevant record or data has been missed. Generally, two contributors screen the search results independently, referring the points of conflict to a third contributor. The same also applies to the data extraction process. The key stages in this step include screening the results, developing a data extraction tool/form/framework, and extraction of data.

Study Selection Study selection usually starts with pooling the records using an aiding tool in order to increase the pace and remove the potential duplications among various databases. Study selection should be based on the pre-defined inclusion and exclusion criteria. Several tools have been proposed to make screening simpler and easier as a team effort, including EndNote, Covidence, and Rayyan (Harrison et al., 2020). Like any other type of SRs, the screening process begins with removing duplicate records, followed by screening by titles, screening by abstracts, and finally screening the remaining records by their full texts. By experience, it might be involving more than one team member could significantly prevent potential biases.

PRISMA Flow Diagram

A complete study flow diagram should be prepared and presented in the ultimate review, displaying the number of records at the beginning and in each stage of the screening process. Figure 1 presents the general template for reporting the DTA-SRs PRISMA flow diagram.

Figure 1. The PRISMA flow diagram template for reporting DTA-SRs

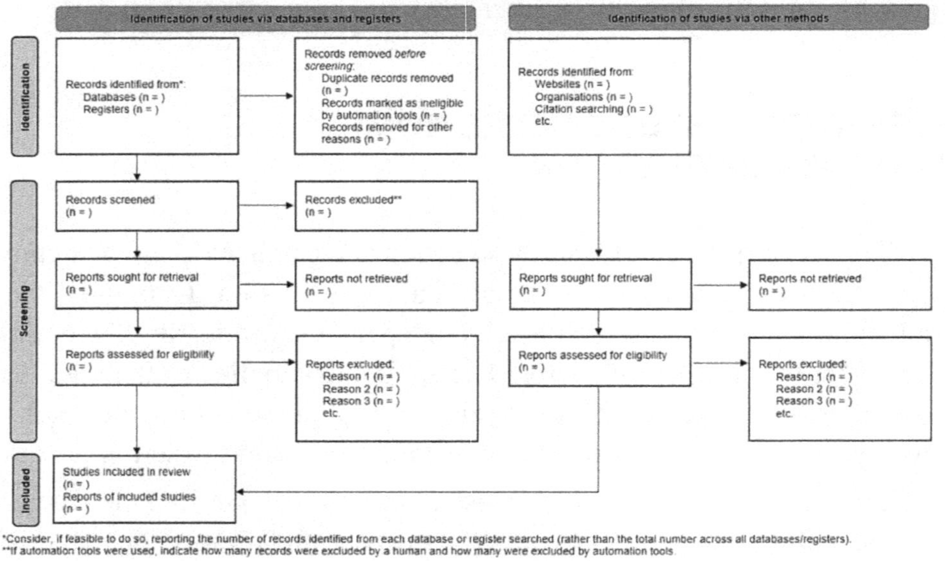

Data Extraction Framework

The principles of data extraction for DTA-SRs are not much different from other SRs. The data extraction framework typically comprises predefined variables and descriptors related to study characteristics, participant demographics, index test(s), gold/reference standard, and diagnostic accuracy measures (such as sensitivity, specificity, and positive/negative predictive values). Standardized data extraction forms have also been introduced to ensure consistency and reproducibility across reviewers.

STEP FOUR: CRITICAL APPRAISAL AND RISK OF BIAS ASSESSMENT

Critical appraisal involves systematically assessing the methodological quality, validity, and reliability of included studies to ascertain the confidence and trustworthiness of their findings. In DTA-SRs, critical appraisal focuses on evaluating the risk of bias, variability, and sources of error in primary diagnostic accuracy studies.

Types of Biases in Diagnostic Studies

All main types of biases, including selection bias, performance bias, detection bias, attrition bias, and reporting bias, and their subtypes, could frequently occur among diagnostic studies. However, some forms of biases are more prevalent among diagnostic studies, including:

Spectrum Bias

As a type of selection bias, spectrum bias arises from systematic differences in the spectrum of disease severity or prevalence between the study population and the target population. Studies conducted in populations with a higher disease prevalence may overestimate test sensitivity and underestimate specificity, while those conducted in populations with lower disease prevalence may result in opposite effects.

Verification Bias and Disease Progression Bias

Verification bias, sometimes referred to as work-up bias, overlaps with disease progression bias – also called index test review bias – which occurs when only a subset of participants undergo verification with the reference standard, when groups within one study are tested with different reference tests, or when the results of the index test influence the decision to pursue a further diagnostic evaluation with the reference standard, leading to differential verification between test-positive and test-negative individuals. This bias arises when diagnostic tests are selectively applied to patients based on clinical suspicion or symptoms, resulting in overestimating test accuracy. Also, this bias occurs when individuals who test positive on the index test are more likely to undergo verification with the reference standard, while those who test negative may not undergo further evaluation, resulting in different definitions or diagnostic thresholds for the diagnosis of interest. Consequently, disease progression bias can lead to an overestimation of test sensitivity and an underestimation of specificity, as individuals with false-negative index test results are less likely to be identified as having the condition of interest.

Incorporation Bias

Incorporation bias occurs when the index test results are used as part of the reference test, or influence the decision to perform the reference standard, thereby inflating estimates of sensitivity and specificity.

Interpretation Bias

Interpretation bias, sometimes referred to as diagnostic review bias or observer bias, occurs when knowledge of the index test result influences the interpretation of the reference standard, leading to differential verification or misclassification of disease status.

Detection Bias

Detection bias can occur when the outcome measures or data collection process differ among the groups. This difference could occur in the frequency of the condition and variations in the modalities of detection/diagnosis, particularly when different reference tests are used among different diagnostic studies. This scope of bias usually overlaps with the verification bias.

Imperfect Standard Bias

If the reference test is far from the diagnostic gold standards, or has been replaced with an alternative reference test (due to economic or ethical restrictions), the imperfect standard bias occurs, usually resulting in an underestimation of the index test.

Tools for Risk of Bias Assessment

Several tools are available for assessing the risk of bias and methodological quality of the studies included in DTA-SRs. These tools provide structured frameworks for evaluating various study conduct and reporting domains. Some commonly used tools for risk of bias assessment in DTA-SR studies are:

QUADAS-2

The Quality Assessment of Diagnostic Accuracy Studies (QUADAS) is a consensus-based tool for quality assessment in DTA studies. QUADAS consists of fourteen items, including various aspects such as patient spectrum, reference standards, disease progression bias, verification bias, incorporation bias, review bias, test execution, study withdrawals, and indeterminate results. Later on, based on experience, anecdotal feedback, and input, areas for improvement were identified, leading to the development of QUADAS-2 (Whiting et al., 2011). This tool consists of four domains: patient selection, index test, reference standard, and flow and timing. Each domain is evaluated for the risk of bias, with the first three also considering concerns about applicability. QUADAS-2 is implemented in different phases, from defining the review question and creating a flowchart for the primary study, to practical bias and applicability evaluation. This tool facilitates a more transparent assessment of bias and applicability in primary diagnostic accuracy research.

Phase 1

Phase 1 involves exploring the review question. During this phase, researchers elaborate on the specific aim of their review, including the target condition, index test, and reference standard. They also identify key aspects of participant selection, such as demographics, comorbidities, and disease severity, ensuring alignment with the intended population outlined in the review question. This phase lays the foundation for the tailored assessment of bias and applicability across the domains of patient selection, index test, reference standard, and flow and timing in subsequent phases of QUADAS-2 application.

Phase 2

In the second phase, a flow diagram for the primary study is drawn to visually represent the selection process of studies included in the review. The flow diagram typically starts with identifying studies through database searches and other sources. Then, it illustrates the screening process, including the number of studies screened, assessed for eligibility, and ultimately included in the review. Any excluded studies and the reasons for their exclusion are also documented in the flow diagram.

Phase 3

Phase 3 involves evaluating each domain of the QUADAS-2 tool – patient selection, index test, reference standard, and flow and timing – using predetermined signaling questions. The initial part of each domain focuses on bias and consists of three sections: information used to assess the risk of bias, signaling questions, and the actual judgment of bias. The aim of documenting the information is to ensure transparency and facilitate discussion among authors independently evaluating assessments. Additional signaling questions are designed to facilitate the judgment. They are answered as "yes," "no," or "unclear" and are phrased such that "yes" indicates a low risk of bias. The risk of bias is assessed as "low," "high," or "unclear." If all signaling questions within a domain receive a "yes" response, bias is judged as low. However, potential bias is identified if any signaling question receives a "no" response. Reviewers must then utilize the guidelines established in phase 2 to assess bias. The "unclear" category is used when insufficient data is available for judgment. Categorized into four main domains, the practical applicability and risk of bias assessment in the third phase of QUADAS-2 comes with a set of signaling questions in each domain, including patient selection, the index test and reference standard, and the flow/timing.

JBI Critical Appraisal Tool

JBI (formerly the Joanna Briggs Institute) is a globally recognized organization promoting evidence-based healthcare practices. With an extensive network and a focus on systematic reviews, JBI provides comprehensive guidance and tools for evaluating research quality and bias. The JBI critical appraisal checklist for DTA studies consists of ten main questions that should be answered as yes/no/unclear/not applicable, covering various aspects, study design, sampling, and randomization, to proper analysis and pre-specified thresholds (Campbell JM, 2020).

ROBIS

Risk of Bias in Systematic Reviews (ROBIS) is a general tool for risk of bias assessment in SRs (Whiting et al., 2016). ROBIS is not merely used for DTA-SRs, but has the applicability for these types of reviews. ROBIS similarly consists of three main phases.

STARD

Standards for Reporting of Diagnostic Accuracy Studies (STARD) is a reporting guideline for primary studies of diagnostic accuracy (Cohen et al., 2016). Although STARD comes with a checklist that could be used in the quality assessment of primary DTA studies, it should be noted that STARD chiefly focuses on the quality of reporting and would probably miss other aspects of the study. Therefore, it might not be the best tool for the risk of bias assessment.

Newcastle-Ottawa scale

Newcastle-Ottawa scale (NOS) was initially introduced to assess the quality of non-randomized studies, categorized into cohort and case-control studies (Wells et al., 2000). NOS includes eight items, categorized into three main categories, including selection, comparability, and exposure/outcome.

How to Report the Assessment Results

Reporting the results of risk of bias assessment in DTA-SRs is essential for transparency and reproducibility. Like any other type of SR, the findings of risk of bias assessment should be clearly and systematically reported to enable readers to evaluate the methodological quality and reliability of included studies. Typically, risk of bias assessment results are presented narratively, summarizing the overall risk of bias for each included study across key domains (e.g., patient selection, index test, reference standard). Additionally, risk of bias tables may be utilized to provide a concise overview of the methodological quality of included studies, categorizing the risk of bias as low, unclear, or high for each domain. Furthermore, graphical representations such as the risk of bias summary figures or bar charts may be employed to visually depict the distribution of risk of bias across studies. Moreover, research could incorporate the results from risk of bias assessment into statistical aspects, stratifying the analysis according to risk of bias or applicability concerns, therefore, restricting inclusion into the review or primary analysis based on results, or using covariates in meta-regression. Regardless of the mean of presentation, detailed explanations and justifications for the assigned risk of bias ratings, including specific concerns or limitations identified within each domain is essential.

STEP FIVE: SYNTHESIS AND ANALYSIS

Synthesis and analysis in DTA-SRs involve the integration of primary diagnostic accuracy data from included studies to provide summary estimates of test performance and assess overall diagnostic accuracy. The synthesis process typically entails pooling sensitivity, specificity, likelihood ratios, and other performance metrics across studies using meta-analytic techniques, such as bivariate or hierarchical modeling. These statistical methods account for the correlation between sensitivity and specificity within studies and yield summary estimates with accompanying measures of uncertainty (such as confidence intervals). Sensitivity analyses may be conducted to explore the robustness of the results. Subgroup analyses may also be performed to further explore sources of heterogeneity and differences in test performance across populations. Additionally, diagnostic accuracy measures may be depicted graphically using forest plots or receiver operating characteristic (ROC) curves to visually represent the distribution of study estimates and summarize overall test performance.

Various approaches are available for synthesizing and presenting the results of SRs, including meta-analysis, narrative synthesis, or a graphical or tabular summary (Cumpston et al., 2023). Likewise, the synthesis in DTA-SRs is basically either qualitative/descriptive or quantitative/statistical. However, all SRs, even the ones with quantitative analysis, also contain narrative and qualitative synthesis.

Qualitative Synthesis

In lack of quantitative data, a qualitative synthesis should be performed. Approaches to systematic review of qualitative research are still evolving, and ongoing debate still persists in this field. Depending on the review question and available data, there are several approaches used for synthesizing qualitative evidence across the SRs, varying from thematic and narrative syntheses to content analysis, meta-ethnography, and meta-aggregation (Flemming & Noyes, 2021). Meta-ethnography involves identifying key concepts from individual studies, translating them into one another, and developing a "line of argument" that goes beyond the content of the original studies. This process of 'going beyond' is considered crucial in synthesis, distinguishing it from traditional literature reviews (France et al., 2019). Meta-aggregation is a protocol-driven synthesis approach with an extensive critical appraisal that aims to aggregate the findings from combined results into categories, and of categories into synthesized findings and conclusions. Among all synthesis approaches, the latter might be considered as the most useful in DTA-SRs. However, most DTA-SRs benefit from quantitative variables and data, requiring further quantitative analyses.

Summary of Findings Table

There is ongoing debate surrounding the synthesis of qualitative research, and on the other hand, due to pooling limits of qualitative evidence, we recommend a summary of findings (SoF) table for qualitative synthesis. While SoF tables are more commonly associated with quantitative evidence synthesis, they can also be adapted for qualitative synthesis in DTA-SRs (Lewin et al., 2018).

In the Cochrane MIF project, an iterative approach was employed to develop SoF tables compatible with diagnostic test studies (*Cochrane Methods GRADEing*). This process involved analyzing data from multiple rounds of discussions, feedback, and user testing, where most participants preferred tabular formats over plain text for summarizing systematic reviews of test accuracy. While users favored less complex tables, they also considered presenting sensitivity and specificity estimates overly simplistic. Consequently, the project led to the development of current SoF table formats for test accuracy, aligning with suggestions from the GRADE approach. These formats can be generated using the GRADEpro Guideline Development Tool (GRADEpro GDT) and are undergoing further refinement to create electronic interactive tables catering to user needs (*GRADEpro GDT*). Generally, a DTA-SR with only qualitative data is not normally expected, and qualitative synthesis usually precedes the meta-analysis.

Meta-Analysis for Quantitative Variables

Synthesizing quantitative data in DTA-SRs includes pooling the primary sensitivity and specificity measures. However, in some stances, other diagnostic measures, such as positive predictive value (PPV) and negative predictive value (NPV), can also be acquired using sensitivity and specificity. Estimation of summary ROC may be conducted depending on the presence or absence of threshold effect.

Meta-analysis involves statistical techniques for synthesizing results from multiple studies, including DTAs. It aggregates findings from various studies to offer average estimates of a test's diagnostic accuracy, along with measures of uncertainty and variability among studies. Numerous methods exist for conducting meta-analyses of DTA studies (Veroniki et al., 2022). DTA-MAs must handle both sensitivity and specificity, along with paying attention to the threshold effect and the trade-off between these two. Therefore, meta-analysis techniques have been developed to combine studies using tests at different thresholds (Steinhauser et al., 2016).

A random-effects model assumes that the included studies represent a random sample from the population of hypothetical studies. This model assumes that there is no singular true effect size, but rather each study's true effect size varies, hence the term "random-effects" (Zhai & Guyatt, 2024). Consequently, even if all studies had infinitely large sample sizes, observed study effects would still vary due to true differences in treatment effects among studies. If a large enough number of studies were performed, the effect estimates from all studies would follow a normal distribution (Liu et al., 2023). The random-effects model, therefore, considers both within-study variability and between-study variability, commonly referred to as heterogeneity. On the other hand, given the expected heterogeneity in DTA study results, random-effects models are more suitable to capture variability across studies (Langan, 2022).

The simple univariate meta-analytic method involves pooling effect size from studies that measure the same outcome of interest. This method focuses on analyzing one outcome variable at a time across studies, typically using weighted averages to combine effect sizes and assess overall trends or associations (van Houwelingen et al., 2002). Although it separately pools sensitivity and specificity, it ignores the threshold effect; on the other hand, this method is more suitable for homogenous studies. Considering these issues, this approach may lead to misleading results, highlighting the fact that it might be better to consider random-effect models rather than fixed-effect models – which are utilized in this method (Takwoingi et al., 2015).

Separate random-effects meta-analyses for negative likelihood ratio (NLR), positive likelihood ratio (PLR), sensitivity, and specificity allow for more precise estimation of individual measures, and authors can tailor their analyses to focus on specific aspects of test performance. However, separating the analysis for each parameter may result in duplication of effort, as researchers need to perform separate analyses for each measure of test performance. It is also associated with the risk of type I error inflation (Brunner & Austin, 2009).

However, bivariate random-effects meta-analysis of ROC curves can address the variability in test thresholds used among studies, allowing for the combination of data from studies using different threshold values. Nevertheless, its complexity and the assumption of bivariate normality limit its usage among researchers (Arends et al., 2008).

The hierarchical summary receiver operating characteristic (HSROC) model is a statistical approach that extends the traditional summary ROC (SROC) model by accommodating both within-study and between-study variability. This model uses a hierarchical structure that enables the estimation of overall summary parameters while accounting for variability within and between studies. This model can estimate summary sensitivity and specificity values, along with their confidence intervals. An advantage of the HSROC model is its ability to handle studies with different thresholds for defining test positivity, or when there is no standard cut-off to define a positive result (Macaskill, 2004). However, it assumes a specific parametric form for the relationship between sensitivity and specificity, which may not always reflect the true underlying distribution in practice. Additionally, the model requires a sufficient number of studies to provide reliable estimates, and its interpretation can be challenging (Dukic & Gatsonis, 2003).

Several software and packages are available for performing meta-analysis on DTA studies, including the SAS macro MetaDAS, the METANDI Stata user-written package (*Software for meta-analysis of DTA studies*), R-packages meta, mada, and diagmeta (Shim, 2022), and Meta-DiSc (Zamora et al., 2006). RevMan also provides limited capacity for DTA-MAs (Wang & Leeflang, 2019).

Subgroup Analysis in DTA Studies

Subgroup analysis is a useful methodological tool for exploring potential sources of heterogeneity and variability in test performance across different patient populations, test methods, or study characteristics. In DTA-SRs, subgroup analysis involves stratifying the included diagnostic accuracy studies based on pre-defined subgroups and examining differences in test performance estimates within each subgroup.

Commonly investigated subgroups in DTA-SRs include patient demographics (such as age, sex, or ethnicity), disease severity, test methods or thresholds, and study quality indicators. Subgroup analyses are typically conducted using statistical techniques such as meta-regression or stratified analysis. Meta-regression allows for exploring the association between study-level characteristics and diagnostic accuracy estimates, while stratified analysis involves calculating summary estimates of test performance within each subgroup separately. However, it is essential to approach subgroup analysis in DTA-SR studies with caution and methodological rigor. Subgroup analyses are recommended be pre-defined in the study protocol or analysis plan to minimize the risk of data-driven results and selective reporting bias. Moreover, the statistical power of subgroup analyses should be carefully considered, as small sample sizes within subgroups may limit the precision and reliability of subgroup-specific estimates.

Subgroup Analysis Based on Study Design

One of the common methods of subgroup analysis in DTA-SRs is based on the design of the included DTA studies, which, unlike other approaches to subgroup analysis, such as age or ethnicity, focuses on the methodological vigor of the results rather than direct clinical implications. In this approach, studies could be categorized into three main categories:

Single-Gate Design

This design involves performing tests in reverse order through prospective data collection. Like population-based case-control studies in epidemiology, sample cases and controls from the same patient population and all patients go through a single set of admission criteria. This design is particularly appropriate for the target populations in which we expect low prevalence of the disease, costly index tests, or potential side effects.

Two-Gate Design With Healthy Controls

In studies where cases and healthy controls are drawn from two distinct populations, two sets of inclusion criteria are applied, one for the diseased and another for the healthy participants. Compared with the single-gate design, it benefits from an improved accuracy estimation but comes with limited generalizability. This increase in accuracy can be attributed to spectrum effects and limited-challenge bias. Including individuals with advanced disease, often the sickest cases, tends to reduce false-negative results, thereby potentially inflating sensitivity estimates. Moreover, using healthy volunteers as controls may increase specificity estimates, as these individuals are less likely to have conditions that could lead to false positives.

Two-Gate Design With Alternative Diagnosis Control

Similar to the latter design, except the control is selected among patients resembling symptoms of the target condition. Depending on the selection of cases, the sensitivity and specificity are greatly affected; overrepresentation of severe or mild disease can skew sensitivity estimates, and specificity may be over- or underestimated depending on the type of alternative diagnosis included.

Heterogeneity in DTA Studies

Unlike systematic reviews of interventions, where similar effect estimates across studies may justify a fixed-effect approach, variability in test accuracy between studies necessitates random-effects meta-analysis methods. Differences in patient characteristics, test methods, study design, and other factors likely contribute to this heterogeneity in test accuracy between studies. Study covariates should preferably be incorporated into hierarchical models to explore heterogeneity efficiently. Meta-regression and subgroup analysis on these covariates could help identify the factors influencing test accuracy (Simmonds & Higgins, 2007).

Subgroup comparisons within studies provide more valid evidence of heterogeneity sources than comparisons between studies. Additionally, Subgroup findings are more credible if they have a scientific rationale, even though such comparisons are rare in systematic reviews due to inadequate reporting of subgroup data. The chance of obtaining significant results increases with the number of statistical comparisons undertaken. However, investigations based on small numbers of studies may not produce useful findings (Dias et al., 2013).

Univariate tests for heterogeneity in sensitivity and specificity, Cochrane Q, Higgins' I^2 statistics are generally discouraged in test accuracy reviews due to their inability to address heterogeneity explained by phenomena such as positivity threshold effects (Lee et al., 2015). Alternative measures, like the variance parameter of random-effects models and multivariate I^2 statistics, have been proposed but are not widely adopted. Multivariate analysis to investigate multiple sources of heterogeneity in parallel is often infeasible due to limited study availability. Graphical displays, such as SROC plots, usually offer a more explicit assessment. Including prediction regions around summary point estimates visually summarizes heterogeneity, accounting for the spread of true underlying test accuracy across studies (Patel et al., 2021).

Reporting Synthesized DTA Results

Results from individual studies should include study identifiers, publication years, total participants, those with the target condition, true/false positives, and true/false negatives, along with measures of statistical uncertainty, typically 95% confidence intervals (95%CIs). These details can be presented in tables or coupled forest plots of sensitivity and specificity. In cases of pairwise comparisons, where multiple estimates of sensitivity and specificity are provided, presenting linked SROC plots showing multiple estimates from each study can help visualize the relationship between pairs of estimates. Authors are encouraged to select relevant and appropriate summary measures for each analysis rather than reporting all possible measures of test accuracy. As discussed, summary estimates commonly include sensitivity and specificity, but other measures like positive and negative predictive values may also be produced, depending on the design of primary studies and the use of different reference standards.

When summarizing estimates from multiple meta-analyses, tables should include the number of studies, participants, and those with and without the target condition for each analysis. However, it is crucial to interpret summary estimates of sensitivity and specificity from meta-analysis as expressions of central tendency across studies, acknowledging uncertainty and variability in accuracy (Trikalinos et al., 2012).

Presentation of SROC curves requires constructions of a 2×2 table containing the counts of true positives, false positives, false negatives, and true negatives from each study, regardless of threshold value, allowing the estimation of an SROC curve. If the SROC curve is assumed to be symmetrical, indicating that accuracy does not depend on the threshold, it can be described by a constant diagnostic odds ratio (DOR) (Takwoingi et al., 2017). The same DOR may result from different combinations of sensitivity and specificity. Single measures like DORs may lack clinical informativeness due to a lack of information on false negatives and false positives. DORs are particularly useful in meta-analysis for comparing tests or subgroups. When two curves are symmetrical, their relative accuracy can be summarized using the ratio of DORs or relative DOR (RDOR) (Glas et al., 2003).

Sensitivity and specificity pairs from the curve can be computed to illustrate test performance. Points on the curve within the sensitivity and specificity estimates range reported by included studies are most informative. Sensitivity and specificity pairs, representing values such as the median and interquartile range from included studies, can be presented in a table for clarity. Values can also be chosen based on context, such as reporting sensitivity at a minimally acceptable specificity level. Values and corresponding estimates can also be organized in a table format for ease of interpretation, along with estimates of likelihood ratios at fixed values.

Authors should convey the degree of statistical uncertainty associated with summary estimates by reporting confidence intervals or credible intervals alongside point estimates (For example, please see (Wei et al., 2016)). Prediction intervals provide insight into heterogeneity by indicating where sensitivity and specificity estimates from future studies are expected to fall. Systematic reviews may include prediction regions with coverage probabilities of 50%, 90%, or 95%, with the 95% prediction regions often covering large areas of ROC space due to substantial heterogeneity. Confidence regions depict uncertainty due to within-study variability, while prediction regions account for both within- and between-study variability.

In cases of high heterogeneity, the 95% prediction region will be much larger than the 95% confidence region. Estimation of prediction intervals relies on assumptions of normal distributions and can be problematic with a small number of studies. Authors should carefully consider the appropriateness of presenting prediction regions when the number of studies is small, typically fewer than ten.

The appropriate statistical measures for comparing tests remain consistent regardless of whether within-study or between-study comparisons are conducted. When comparing two tests, focus on describing the magnitude and direction of the difference, uncertainty in estimates, and degree of heterogeneity. Comparing summary sensitivity and specificity estimates for each test along with measures of absolute or relative differences illustrates the potential impact of using one test over another. If SROC curves have the same shape, the RDOR is constant along the curve, facilitating comparison. Overlapping confidence intervals do not necessarily indicate the lack of statistical significance; P values are needed for this determination.

STEP SIX: CLINICAL UTILITY AND IMPLICATIONS FOR DECISION-MAKING

The ultimate goal of DTA-SRs is to inform clinical decision-making by providing evidence on the performance of diagnostic tests. However, translating diagnostic accuracy into clinical utility requires careful consideration of how test results affect patient outcomes and general healthcare decisions.

Translating Results to Clinical Utility

Clinical utility refers to the ability of a diagnostic test to improve patient outcomes. It includes not only the accuracy of the test itself, but also its impact on clinical decisions, patient management, and overall health benefits. Translating the results of DTA-SRs into clinical utility involves several steps:

Interpretation of Test Accuracy

Clinicians must understand the implications of sensitivity, specificity, and predictive values in the context of the patient's clinical scenario.

Risk-Benefit Analysis

The potential benefits of an accurate diagnosis must be weighed against the risks of false positives or negatives, which could lead to unnecessary interventions or missed treatments.

Outcome Relevance

Tests should provide information leading to interventions that improve patient outcomes, such as reduced morbidity or mortality.

Feasibility and Accessibility

The tests must be practical and available within the healthcare setting where the patient is being treated.

Frameworks for Evaluating Clinical Utility

Several frameworks are available to evaluate the clinical utility of diagnostic tests. These frameworks guide clinicians and policymakers in determining whether a test should be incorporated into practice:

GRADE Approach

The grading of recommendations assessment, development, and evaluation (GRADE) approach assesses the quality of evidence and the strength of recommendations, including diagnostic tests (Gopalakrishna et al., 2014).

Fryback and Thornbury's Hierarchy of Efficacy

This model outlines six levels of efficacy, from technical quality to societal efficacy, providing a comprehensive view of a test's impact (Fryback & Thornbury, 1991).

Wilson and Jungner Criteria

Initially developed for screening programs, these criteria could also be applied to diagnostic tests, focusing on the importance, treatability, and cost-effectiveness of the condition being tested for (Andermann et al., 2008).

ACCE model

Consisting of analytic validity, clinical validity, clinical utility, and ethical, legal and social implications, this framework specifically addresses the components necessary to evaluate the utility and broader implications of genetic tests (National Academies of Sciences, 2017).

Patient Preferences and Values in Diagnostic Decision-Making

Patient-centered care is a fundamental aspect of modern healthcare, and diagnostic decision-making is no exception. Patient preferences and values play a vital role in determining the clinical utility of a diagnostic test:

Understanding Patient Context

DTA-SR conclusions should be drawn considering the fact that each patient has unique clinical and personal circumstances, including comorbidities, risk factors, and social support, which must be factored into the ultimate decision about whether to use a diagnostic test.

Communication of Risks and Benefits

Patients should be informed about the potential outcomes of a test in a way that is understandable and relevant to their situation. This includes discussing the likelihood of false positives or negatives and the implications of such results. Therefore, DTA-SRs are encouraged to provide as much details as possible in this regard.

Shared Decision-Making

As a pillar of evidence-based medicine, patients should be engaged in discussions about potential diagnostic tests to value and consider their preferences. This collaborative process leads to personalized care and improved patient satisfaction.

Respect for Autonomy

Even in the presence of the highest levels of evidence, individuals have the right to refuse a diagnostic test. The patients' right to make informed choices about their healthcare, including the acceptance or refusal of diagnostic tests, must be respected by clinicians and policymakers.

Overall, the clinical utility of diagnostic tests and the implications for decision-making are complicated. DTA-SRs provide valuable information on test accuracy, but translating these results into clinical practice requires a thorough evaluation of clinical utility, considering frameworks that assess this utility, and integrating patient preferences and values.

ADDITIONAL TOPICS AND EXTENSIONS

Guidelines for Reporting Diagnostic Reviews

Guidelines for reporting diagnostic reviews are essential to ensure transparency, consistency, and completeness in reporting methodological processes and findings. The Preferred Reporting Items for Systematic

Reviews and Meta-Analyses (PRISMA) statement provides a comprehensive guideline for reporting systematic reviews, including those focused on diagnostic accuracy (Page et al., 2021). The PRISMA-DTA extension specifically addresses reporting standards for DTA-SRs (McInnes et al., 2018). Additionally, adjunctive compliance with other reporting guidelines, such as the Standards for Reporting of Diagnostic Accuracy Studies (STARD), ensures comprehensive and transparent reporting of primary diagnostic accuracy studies (Cohen et al., 2016), thereby facilitating the integration of study findings into systematic reviews and meta-analyses. Additional guidelines, such as the transparent reporting of multivariable prediction models for individual prognosis or diagnosis (TRIPOD) and its extension for systematic reviews and meta-analyses (TRIPOD-SRMA) are also available, but limited to their very specific uses (Collins et al., 2015; Snell et al., 2023).

Figure 2 provides the PRISMA-DTA checklist (*PRISMA for Diagnostic Test Accuracy*). Please note that PRISMA-DTA also provides recommendations regarding the abstract section of DTA-SRs (accessible from https://www.prisma-statement.org/dta).

Cochrane Reviews of Diagnostic Test Accuracy

Cochrane reviews of DTA studies adhere to the established Cochrane methodology to ensure the validity and reliability of the results. Specifically, Cochrane DTA reviews adhere to the Cochrane Handbook for Diagnostic Test Accuracy Reviews, which provides detailed guidance on conducting DTA reviews within the Cochrane framework (Deeks et al., 2023). Key requirements include adherence to PRISMA-DTA reporting guidelines, utilization of validated tools for risk of bias assessment (QUADAS-2), and rigorous statistical analysis of diagnostic accuracy data. Table 2 presents the full outline and structure of a Cochrane DTA review. The complete template for Cochrane protocols and reports of DTA studies is available in the Cochrane Handbook for Diagnostic Test Accuracy Reviews.

Table 2. The complete outline of Cochrane DTA reviews

Sections	Specific components
Title	–
Abstract	–
Plain language summary	–
Summary of findings	–
Background	• Target condition being diagnosed • Index tests • Clinical pathway – Prior tests – Role of index tests – Alternative tests • Rationale
Objectives	–
Methods	• Criteria for considering studies for this review – Types of studies – Participants – Index tests – Target conditions – Reference standards • Search methods for identification of studies – Electronic searches – Searching other resources • Data collection and analysis – Selection of studies – Data extraction and management – Assessment of methodological quality – Statistical analysis and data synthesis – Investigations of heterogeneity – Sensitivity analyses – Assessment of reporting bias
Results	• Results of search • Methodological quality of included studies • Findings
Discussion	• Summary of main results • Strengths and weaknesses of review • Applicability of findings to review question
Conclusions	• Implications for practice • Implications for research
Appendices	• Search strategy • Review-specific tailoring of QUADAS2

continued on following page

Table 2. Continued

Sections	Specific components
Information	• Authors • Authors' contribution • Sources of support • Declarations of interest • Acknowledgements • Version history • Differences between protocol and review
References	–
Characteristics of studies	–
Data and analyses	–
Figures and tables	–

CONCLUSION AND CLOSING

In conclusion, DTA-SRs play a pivotal role in synthesizing evidence on the performance of diagnostic tests, providing clinicians, policymakers, and researchers with valuable insights into test accuracy and reliability. Through rigorous methodological approaches, DTA-SRs enable the integration of diverse diagnostic accuracy data into evidence-based clinical practice decision-making. The established reporting guidelines and validated tools make a clear path for researchers aiming to conduct DTA-SRs. However, DTA-SRs usually need more patience, precision, and time dedication than other SRs. Moving forward, continued advancements in methodological approaches, innovative statistical techniques, and interdisciplinary collaborations will further enhance the quality and utility of DTA-SRs, ultimately improving diagnostic decision-making and patient care outcomes.

REFERENCES

About: searchRxiv. Retrieved May 9 from https://www.cabidigitallibrary.org/journal/searchrxiv/about

About PROSPERO. https://www.crd.york.ac.uk/prospero/#aboutpage

About the Research Registry. Retrieved May 9 from https://www.researchregistry.com/about

About us - protocols.io. Retrieved May 9 from https://www.protocols.io/about

Andermann, A., Blancquaert, I., Beauchamp, S., & Déry, V. (2008). Revisiting Wilson and Jungner in the genomic age: A review of screening criteria over the past 40 years. *Bulletin of the World Health Organization*, *86*(4), 317–319. DOI: 10.2471/BLT.07.050112 PMID: 18438522

Arends, L. R., Hamza, T. H., van Houwelingen, J. C., Heijenbrok-Kal, M. H., Hunink, M. G. M., & Stijnen, T. (2008). Bivariate Random Effects Meta-Analysis of ROC Curves. *Medical Decision Making*, *28*(5), 621–638. DOI: 10.1177/0272989X08319957 PMID: 18591542

Behringer, W., & Freund, Y. (2021). Clinical translation of diagnostic studies: Pitfalls of the usual reported characteristics. *European Journal of Emergency Medicine*, *28*(3), 165–166. DOI: 10.1097/MEJ.0000000000000830 PMID: 33899776

Bernardo, W. M. (2017). PRISMA statement and PROSPERO. *International Braz J Urol*, *43*(3), 383–384. DOI: 10.1590/s1677-5538.ibju.2017.03.02 PMID: 28520335

Brunner, J., & Austin, P. C. (2009). Inflation of Type I error rate in multiple regression when independent variables are measured with error. *The Canadian Journal of Statistics*, *37*(1), 33–46. https://doi.org/https://doi.org/10.1002/cjs.10004. DOI: 10.1002/cjs.10004

Campbell, J. M. K. M., Ding S, Carmody DP, Hakonsen SJ, Jadotte YT, White S, Munn Z. (2020). *Chapter 9: Diagnostic test accuracy systematic reviews.* JBI. *Cochrane Methods GRADEing*. Retrieved May 9 from https://methods.cochrane.org/gradeing/research

Cohen, J. F., Korevaar, D. A., Altman, D. G., Bruns, D. E., Gatsonis, C. A., Hooft, L., Irwig, L., Levine, D., Reitsma, J. B., de Vet, H. C. W., & Bossuyt, P. M. (2016). STARD 2015 guidelines for reporting diagnostic accuracy studies: Explanation and elaboration. *BMJ Open*, *6*(11), e012799. DOI: 10.1136/bmjopen-2016-012799 PMID: 28137831

Collins, G. S., Reitsma, J. B., Altman, D. G., & Moons, K. G. M. (2015). Transparent reporting of a multivariable prediction model for individual prognosis or diagnosis (TRIPOD): The TRIPOD Statement. *BMC Medicine*, *13*(1), 1. DOI: 10.1186/s12916-014-0241-z PMID: 25563062

Cumpston, M. S., Brennan, S. E., Ryan, R., & McKenzie, J. E. (2023). Synthesis methods other than meta-analysis were commonly used but seldom specified: Survey of systematic reviews. *Journal of Clinical Epidemiology*, *156*, 42–52. DOI: 10.1016/j.jclinepi.2023.02.003 PMID: 36758885

Deeks, J. J., Bossuyt, P. M., Leeflang, M. M., & Takwoingi, Y. (2023). *Cochrane handbook for systematic reviews of diagnostic test accuracy*. John Wiley & Sons. DOI: 10.1002/9781119756194

DeMars, M. M., & Perruso, C. (2022). MeSH and text-word search strategies: Precision, recall, and their implications for library instruction. *Journal of the Medical Library Association: JMLA*, *110*(1), 23–33. DOI: 10.5195/jmla.2022.1283 PMID: 35210959

Dias, S., Sutton, A. J., Welton, N. J., & Ades, A. E. (2013). Evidence Synthesis for Decision Making 3: Heterogeneity—Subgroups, Meta-Regression, Bias, and Bias-Adjustment. *Medical Decision Making*, *33*(5), 618–640. DOI: 10.1177/0272989X13485157 PMID: 23804507

Dukic, V., & Gatsonis, C. (2003). Meta-analysis of diagnostic test accuracy assessment studies with varying number of thresholds. *Biometrics*, *59*(4), 936–946. DOI: 10.1111/j.0006-341X.2003.00108.x PMID: 14969472

Ewald, H., Klerings, I., Wagner, G., Heise, T. L., Stratil, J. M., Lhachimi, S. K., Hemkens, L. G., Gartlehner, G., Armijo-Olivo, S., & Nussbaumer-Streit, B. (2022). Searching two or more databases decreased the risk of missing relevant studies: A metaresearch study. *Journal of Clinical Epidemiology*, *149*, 154–164. https://doi.org/https://doi.org/10.1016/j.jclinepi.2022.05.022. DOI: 10.1016/j.jclinepi.2022.05.022 PMID: 35654269

Flemming, K., & Noyes, J. (2021). Qualitative evidence synthesis: Where are we at? *International Journal of Qualitative Methods*, *20*, 1609406921993276. DOI: 10.1177/1609406921993276

France, E. F., Uny, I., Ring, N., Turley, R. L., Maxwell, M., Duncan, E. A. S., Jepson, R. G., Roberts, R. J., & Noyes, J. (2019). A methodological systematic review of meta-ethnography conduct to articulate the complex analytical phases. *BMC Medical Research Methodology*, *19*(1), 35. DOI: 10.1186/s12874-019-0670-7 PMID: 30777031

Fryback, D. G., & Thornbury, J. R. (1991). The efficacy of diagnostic imaging. *Medical Decision Making*, *11*(2), 88–94. DOI: 10.1177/0272989X9101100203 PMID: 1907710

Fu, H., Wang, L., Zhang, W., Lu, J., & Yang, M. (2023). Diagnostic test accuracy of ultrasound for sarcopenia diagnosis: A systematic review and meta-analysis. *Journal of Cachexia, Sarcopenia and Muscle*, *14*(1), 57–70. DOI: 10.1002/jcsm.13149 PMID: 36513380

Glanville, J., Kotas, E., Featherstone, R., & Dooley, G. (2020). Which are the most sensitive search filters to identify randomized controlled trials in MEDLINE? *Journal of the Medical Library Association: JMLA*, *108*(4), 556–563. DOI: 10.5195/jmla.2020.912 PMID: 33013212

Glas, A. S., Lijmer, J. G., Prins, M. H., Bonsel, G. J., & Bossuyt, P. M. (2003). The diagnostic odds ratio: A single indicator of test performance. *Journal of Clinical Epidemiology*, *56*(11), 1129–1135. DOI: 10.1016/S0895-4356(03)00177-X PMID: 14615004

Gopalakrishna, G., Mustafa, R. A., Davenport, C., Scholten, R. J., Hyde, C., Brozek, J., Schünemann, H. J., Bossuyt, P. M. M., Leeflang, M. M. G., & Langendam, M. W. (2014). Applying Grading of Recommendations Assessment, Development and Evaluation (GRADE) to diagnostic tests was challenging but doable. *Journal of Clinical Epidemiology*, *67*(7), 760–768. DOI: 10.1016/j.jclinepi.2014.01.006 PMID: 24725643

GRADEpro GDT. https://www.gradepro.org/

Harrison, H., Griffin, S. J., Kuhn, I., & Usher-Smith, J. A. (2020). Software tools to support title and abstract screening for systematic reviews in healthcare: An evaluation. *BMC Medical Research Methodology*, *20*(1), 7. DOI: 10.1186/s12874-020-0897-3 PMID: 31931747

Hosseini, M.-S., Jahanshahlou, F., Akbarzadeh, M. A., Zarei, M., & Vaez-Gharamaleki, Y. (2024). Formulating research questions for evidence-based studies. *Journal of Medicine, Surgery, and Public Health*, *2*, 100046. https://doi.org/https://doi.org/10.1016/j.glmedi.2023.100046. DOI: 10.1016/j.glmedi.2023.100046

Langan, D. (2022). Assessing HeterogeneityHeterogeneity in Random-Effects Meta-analysis. In Evangelou, E., & Veroniki, A. A. (Eds.), *Meta-Research: Methods and Protocols* (pp. 67–89). Springer US., DOI: 10.1007/978-1-0716-1566-9_4

Learn more about INPLASY. Retrieved May 9 from https://inplasy.com/learn-more-about-inplasy/

Lee, J., Kim, K. W., Choi, S. H., Huh, J., & Park, S. H. (2015). Systematic Review and Meta-Analysis of Studies Evaluating Diagnostic Test Accuracy: A Practical Review for Clinical Researchers-Part II. Statistical Methods of Meta-Analysis. *Korean Journal of Radiology*, *16*(6), 1188–1196. DOI: 10.3348/kjr.2015.16.6.1188 PMID: 26576107

Leeflang, M. M. (2014). Systematic reviews and meta-analyses of diagnostic test accuracy. *Clinical Microbiology and Infection*, *20*(2), 105–113. DOI: 10.1111/1469-0691.12474 PMID: 24274632

Lewin, S., Bohren, M., Rashidian, A., Munthe-Kaas, H., Glenton, C., Colvin, C. J., Garside, R., Noyes, J., Booth, A., Tunçalp, Ö., Wainwright, M., Flottorp, S., Tucker, J. D., & Carlsen, B. (2018). Applying GRADE-CERQual to qualitative evidence synthesis findings—paper 2: How to make an overall CERQual assessment of confidence and create a Summary of Qualitative Findings table. *Implementation Science : IS*, *13*(1), 10. DOI: 10.1186/s13012-017-0689-2 PMID: 29384082

Liu, Z., Al Amer, F. M., Xiao, M., Xu, C., Furuya-Kanamori, L., Hong, H., Siegel, L., & Lin, L. (2023). The normality assumption on between-study random effects was questionable in a considerable number of Cochrane meta-analyses. *BMC Medicine*, *21*(1), 112. DOI: 10.1186/s12916-023-02823-9 PMID: 36978059

Liu, Z., Yao, Z., Li, C., Liu, X., Chen, H., & Gao, C. (2013). A step-by-step guide to the systematic review and meta-analysis of diagnostic and prognostic test accuracy evaluations. *British Journal of Cancer*, *108*(11), 2299–2303. DOI: 10.1038/bjc.2013.185 PMID: 23695015

Macaskill, P. (2004). Empirical Bayes estimates generated in a hierarchical summary ROC analysis agreed closely with those of a full Bayesian analysis. *Journal of Clinical Epidemiology*, *57*(9), 925–932. DOI: 10.1016/j.jclinepi.2003.12.019 PMID: 15504635

Mathew, M. J. (2024). Literature search in systematic reviews: How much is good enough? *Clinical Epidemiology and Global Health*, *25*, 101485. https://doi.org/https://doi.org/10.1016/j.cegh.2023.101485. DOI: 10.1016/j.cegh.2023.101485

McInnes, M. D. F., Moher, D., Thombs, B. D., McGrath, T. A., Bossuyt, P. M., Clifford, T., Cohen, J. F., Deeks, J. J., Gatsonis, C., Hooft, L., Hunt, H. A., Hyde, C. J., Korevaar, D. A., Leeflang, M. M. G., Macaskill, P., Reitsma, J. B., Rodin, R., Rutjes, A. W. S., Salameh, J.-P., & Willis, B. H. (2018). Preferred Reporting Items for a Systematic Review and Meta-analysis of Diagnostic Test Accuracy Studies: The PRISMA-DTA Statement. *Journal of the American Medical Association*, *319*(4), 388–396. DOI: 10.1001/jama.2017.19163 PMID: 29362800

National Academies of Sciences. E., and Medicine; Health and Medicine Division; Board on the Health of Select Populations; Board on Health Care Services; Committee on the Evidence Base for Genetic Testing. (2017). *An Evidence Framework for Genetic Testing*. National Academies Press (US). https://doi.org/DOI: 10.17226/24632

Olsen, M., Zhelev, Z., Hunt, H., Peters, J. L., Bossuyt, P., & Hyde, C. (2019). Use of test accuracy study design labels in NICE's diagnostic guidance. *Diagnostic and Prognostic Research*, *3*(1), 17. DOI: 10.1186/s41512-019-0062-9 PMID: 31517065

Page, M. J., McKenzie, J. E., Bossuyt, P. M., Boutron, I., Hoffmann, T. C., Mulrow, C. D., Shamseer, L., Tetzlaff, J. M., Akl, E. A., Brennan, S. E., Chou, R., Glanville, J., Grimshaw, J. M., Hróbjartsson, A., Lalu, M. M., Li, T., Loder, E. W., Mayo-Wilson, E., McDonald, S., & Moher, D. (2021). The PRISMA 2020 statement: An updated guideline for reporting systematic reviews. *BMJ (Clinical Research Ed.)*, *372*(71), n71. Advance online publication. DOI: 10.1136/bmj.n71 PMID: 33782057

Patel, A., Cooper, N., Freeman, S., & Sutton, A. (2021). Graphical enhancements to summary receiver operating characteristic plots to facilitate the analysis and reporting of meta-analysis of diagnostic test accuracy data. *Research Synthesis Methods*, *12*(1), 34–44. DOI: 10.1002/jrsm.1439 PMID: 32706182

Pieper, D., & Rombey, T. (2022). Where to prospectively register a systematic review. *Systematic Reviews*, *11*(1), 8. DOI: 10.1186/s13643-021-01877-1 PMID: 34998432

PRISMA for Diagnostic Test Accuracy. Retrieved May 9 from https://www.prisma-statement.org/dta

Rutjes, A. W., Reitsma, J. B., Vandenbroucke, J. P., Glas, A. S., & Bossuyt, P. M. (2005). Case-control and two-gate designs in diagnostic accuracy studies. *Clinical Chemistry*, *51*(8), 1335–1341. DOI: 10.1373/clinchem.2005.048595 PMID: 15961549

Salameh, J.-P., Bossuyt, P. M., McGrath, T. A., Thombs, B. D., Hyde, C. J., Macaskill, P., Deeks, J. J., Leeflang, M., Korevaar, D. A., Whiting, P., Takwoingi, Y., Reitsma, J. B., Cohen, J. F., Frank, R. A., Hunt, H. A., Hooft, L., Rutjes, A. W. S., Willis, B. H., Gatsonis, C., & McInnes, M. D. F. (2020). Preferred reporting items for systematic review and meta-analysis of diagnostic test accuracy studies (PRISMA-DTA): Explanation, elaboration, and checklist. *BMJ (Clinical Research Ed.)*, *370*, m2632. DOI: 10.1136/bmj.m2632 PMID: 32816740

Shim, S. R. (2022). Meta-analysis of diagnostic test accuracy studies with multiple thresholds for data integration. *Epidemiology and Health*, *44*, e2022083. DOI: 10.4178/epih.e2022083 PMID: 36228672

Simmonds, M. C., & Higgins, J. P. T. (2007). Covariate heterogeneity in meta-analysis: Criteria for deciding between meta-regression and individual patient data. *Statistics in Medicine*, *26*(15), 2982–2999. https://doi.org/https://doi.org/10.1002/sim.2768. DOI: 10.1002/sim.2768 PMID: 17195960

Snell, K. I. E., Levis, B., Damen, J. A. A., Dhiman, P., Debray, T. P. A., Hooft, L., Reitsma, J. B., Moons, K. G. M., Collins, G. S., & Riley, R. D. (2023). Transparent reporting of multivariable prediction models for individual prognosis or diagnosis: Checklist for systematic reviews and meta-analyses (TRIPOD-SRMA). *BMJ (Clinical Research Ed.)*, *381*, e073538. DOI: 10.1136/bmj-2022-073538 PMID: 37137496

Software for meta-analysis of DTA studies. https://methods.cochrane.org/sdt/software-meta-analysis-dta-studies

Steinhauser, S., Schumacher, M., & Rücker, G. (2016). Modelling multiple thresholds in meta-analysis of diagnostic test accuracy studies. *BMC Medical Research Methodology*, *16*(1), 97. DOI: 10.1186/s12874-016-0196-1 PMID: 27520527

Takwoingi, Y., Guo, B., Riley, R. D., & Deeks, J. J. (2017). Performance of methods for meta-analysis of diagnostic test accuracy with few studies or sparse data. *Statistical Methods in Medical Research*, *26*(4), 1896–1911. DOI: 10.1177/0962280215592269 PMID: 26116616

Takwoingi, Y., & Quinn, T. J. (2018). Review of Diagnostic Test Accuracy (DTA) studies in older people. *Age and Ageing*, *47*(3), 349–355. DOI: 10.1093/ageing/afy023 PMID: 29528366

Takwoingi, Y., Riley, R. D., & Deeks, J. J. (2015). Meta-analysis of diagnostic accuracy studies in mental health. *Evidence-Based Mental Health*, *18*(4), 103–109. DOI: 10.1136/eb-2015-102228 PMID: 26446042

The Open Science Framework. Retrieved May 9 from https://osf.io/

Trikalinos, T. A., Balion, C. M., Coleman, C. I., Griffith, L., Santaguida, P. L., Vandermeer, B., & Fu, R. (2012). Chapter 8: meta-analysis of test performance when there is a "gold standard". *J Gen Intern Med, 27 Suppl 1*(Suppl 1), S56-66. https://doi.org/DOI: 10.1007/s11606-012-2029-1

van der Braak, K., Heus, P., Orelio, C., Netterström-Wedin, F., Robinson, K. A., Lund, H., & Hooft, L. (2023). Perspectives on systematic review protocol registration: A survey amongst stakeholders in the clinical research publication process. *Systematic Reviews*, *12*(1), 234. DOI: 10.1186/s13643-023-02405-z PMID: 38098085

van Houwelingen, H. C., Arends, L. R., & Stijnen, T. (2002). Advanced methods in meta-analysis: Multivariate approach and meta-regression. *Statistics in Medicine*, *21*(4), 589–624. DOI: 10.1002/sim.1040 PMID: 11836738

Veroniki, A. A., Tsokani, S., Agarwal, R., Pagkalidou, E., Rücker, G., Mavridis, D., & Takwoingi, Y. (2022). Diagnostic test accuracy network meta-analysis methods: A scoping review and empirical assessment. *Journal of Clinical Epidemiology*, *146*, 86–96. https://doi.org/https://doi.org/10.1016/j.jclinepi.2022.02.001. DOI: 10.1016/j.jclinepi.2022.02.001 PMID: 35181490

Wang, J., & Leeflang, M. (2019). Recommended software/packages for meta-analysis of diagnostic accuracy. *Journal of Laboratory and Precision Medicine*, *4*, 4. DOI: 10.21037/jlpm.2019.06.01

Wei, T.-T., Hu, Z.-D., Qin, B.-D., Ma, N., Tang, Q.-Q., Wang, L.-L., Zhou, L., & Zhong, R.-Q. (2016). Diagnostic Accuracy of Procalcitonin in Bacterial Meningitis Versus Nonbacterial Meningitis: A Systematic Review and Meta-Analysis. *Medicine*, *95*(11), e3079. https://journals.lww.com/md-journal/fulltext/2016/03150/diagnostic_accuracy_of_procalcitonin_in_bacterial.47.aspx. DOI: 10.1097/MD.0000000000003079 PMID: 26986140

Wells, G. A., Shea, B., O'Connell, D., Peterson, J., Welch, V., Losos, M., & Tugwell, P. (2000). The Newcastle-Ottawa Scale (NOS) for assessing the quality of nonrandomised studies in meta-analyses.

Whiting, P., Savović, J., Higgins, J. P., Caldwell, D. M., Reeves, B. C., Shea, B., Davies, P., Kleijnen, J., & Churchill, R. (2016). ROBIS: A new tool to assess risk of bias in systematic reviews was developed. *Journal of Clinical Epidemiology*, *69*, 225–234. DOI: 10.1016/j.jclinepi.2015.06.005 PMID: 26092286

Whiting, P. F., Rutjes, A. W., Westwood, M. E., Mallett, S., Deeks, J. J., Reitsma, J. B., Leeflang, M. M. G., Sterne, J. A. C., & Bossuyt, P. M. (2011). QUADAS-2: A revised tool for the quality assessment of diagnostic accuracy studies. *Annals of Internal Medicine*, *155*(8), 529–536. DOI: 10.7326/0003-4819-155-8-201110180-00009 PMID: 22007046

Willis, L. D. (2023). Formulating the Research Question and Framing the Hypothesis. *Respiratory Care*, *68*(8), 1180–1185. DOI: 10.4187/respcare.10975 PMID: 37041024

Zamora, J., Abraira, V., Muriel, A., Khan, K., & Coomarasamy, A. (2006). Meta-DiSc: A software for meta-analysis of test accuracy data. *BMC Medical Research Methodology*, *6*(1), 31. DOI: 10.1186/1471-2288-6-31 PMID: 16836745

Zhai, C., & Guyatt, G. (2024). Fixed-effect and random-effects models in meta-analysis. *Chinese Medical Journal*, *137*(1), 1–4. DOI: 10.1097/CM9.0000000000002814 PMID: 37612263

ADDITIONAL READING

Campbell, J. M. K. M., Ding S, Carmody DP, Hakonsen SJ, Jadotte YT, White S, Munn Z. (2020). Chapter 9: Diagnostic test accuracy systematic reviews. JBI.

Campbell, J. M., Klugar, M., Ding, S., Carmody, D. P., Hakonsen, S. J., Jadotte, Y. T., White, S., & Munn, Z. (2015). Diagnostic test accuracy: Methods for systematic review and meta-analysis. *International Journal of Evidence-Based Healthcare*, *13*(3), 154–162. DOI: 10.1097/XEB.0000000000000061 PMID: 26355602

Cohen, J. F., Deeks, J. J., Hooft, L., Salameh, J. P., Korevaar, D. A., Gatsonis, C., Hopewell, S., Hunt, H. A., Hyde, C. J., Leeflang, M. M., Macaskill, P., McGrath, T. A., Moher, D., Reitsma, J. B., Rutjes, A. W. S., Takwoingi, Y., Tonelli, M., Whiting, P., Willis, B. H., & McInnes, M. D. F. (2021). Preferred reporting items for journal and conference abstracts of systematic reviews and meta-analyses of diagnostic test accuracy studies (PRISMA-DTA for Abstracts): Checklist, explanation, and elaboration. *BMJ (Clinical Research Ed.)*, *372*(265), n265. DOI: 10.1136/bmj.n265 PMID: 33722791

Deeks, J. J., Bossuyt, P. M., Leeflang, M. M., & Takwoingi, Y. (2023). *Cochrane handbook for systematic reviews of diagnostic test accuracy*. John Wiley & Sons. DOI: 10.1002/9781119756194

Hosseini, M. S., Jahanshahlou, F., Akbarzadeh, M. A., Zarei, M., & Vaez-Gharamaleki, Y. (2024). Formulating research questions for evidence-based studies. *Journal of Medicine, Surgery, and Public Health*, *2*, 100046. DOI: 10.1016/j.glmedi.2023.100046

Leeflang, M. M. (2014). Systematic reviews and meta-analyses of diagnostic test accuracy. *Clinical Microbiology and Infection*, *20*(2), 105–113. DOI: 10.1111/1469-0691.12474 PMID: 24274632

McInnes, M. D. F., Moher, D., Thombs, B. D., McGrath, T. A., Bossuyt, P. M., Clifford, T., Cohen, J. F., Deeks, J. J., Gatsonis, C., Hooft, L., Hunt, H. A., Hyde, C. J., Korevaar, D. A., Leeflang, M. M. G., Macaskill, P., Reitsma, J. B., Rodin, R., Rutjes, A. W. S., Salameh, J.-P., & Willis, B. H.PRISMA-DTA Group. (2018). Preferred Reporting Items for a Systematic Review and Meta-analysis of Diagnostic Test Accuracy Studies: The PRISMA-DTA Statement. *Journal of the American Medical Association, 319*(4), 388–396. DOI: 10.1001/jama.2017.19163 PMID: 29362800

Page, M. J., McKenzie, J. E., Bossuyt, P. M., Boutron, I., Hoffmann, T. C., & Mulrow, C. D., (2021). The PRISMA 2020 statement: An updated guideline for reporting systematic reviews. *BMJ (Clinical Research Ed.), 372*(71). PMID: 33782057

Rutjes, A. W., Reitsma, J. B., Vandenbroucke, J. P., Glas, A. S., & Bossuyt, P. M. (2005). Case-control and two-gate designs in diagnostic accuracy studies. *Clinical Chemistry, 51*(8), 1335–1341. DOI: 10.1373/clinchem.2005.048595 PMID: 15961549

Salameh, J. P., Bossuyt, P. M., McGrath, T. A., Thombs, B. D., Hyde, C. J., Macaskill, P., Deeks, J. J., Leeflang, M., Korevaar, D. A., Whiting, P., Takwoingi, Y., Reitsma, J. B., Cohen, J. F., Frank, R. A., Hunt, H. A., Hooft, L., Rutjes, A. W. S., Willis, B. H., Gatsonis, C., & McInnes, M. D. F. (2020). Preferred reporting items for systematic review and meta-analysis of diagnostic test accuracy studies (PRISMA-DTA): Explanation, elaboration, and checklist. *BMJ (Clinical Research Ed.), 370*, m2632. DOI: 10.1136/bmj.m2632 PMID: 32816740

KEY TERMS AND DEFINITIONS

Diagnostic Odds Ratio (DOR): The diagnostic odds ratio measures a diagnostic test's overall performance, considering both sensitivity and specificity. It is calculated as the ratio of the odds of a true positive result to the odds of a false positive result.

Diagnostic Test Accuracy Studies: DTA studies evaluate the ability of a proposed test (called index test) to detect individuals with a pre-defined particular disease or condition, estimating the sensitivity and specificity of the index test.

Gold Standard: The gold standard refers to the most accurate and reliable diagnostic or measurement tool available for a particular disease or condition.

Index Test: An index test is a diagnostic test being evaluated to determine its accuracy or effectiveness in identifying a particular disease or condition.

MeSH Term: Medical Subject Headings (MeSH) term is a standardized vocabulary used by the National Library of Medicine to index and categorize biomedical information in databases. MeSH terms help to improve the accuracy and efficiency of literature searches.

Negative Predictive Value (NPV): The negative predictive value is the proportion of true negative results among all individuals who test negative on a diagnostic test. It indicates the probability that a negative test result is correct.

Positive Predictive Value (PPV): The positive predictive value is the proportion of true positive results among all individuals who test positive on a diagnostic test. It indicates the probability that a positive test result is correct.

Receiver Operating Characteristic (ROC) Curve: The ROC curve is a graphical representation of the trade-off between sensitivity and specificity for a diagnostic test across different threshold values. It is used to evaluate and compare the performance of different tests.

Reference Test: The reference test is a diagnostic test that is considered to be the gold standard or has generally acceptable accuracy in diagnosing a particular condition; therefore, it is used to validate the results of other (index) tests.

Search Recall: Search recall, also known as search sensitivity, refers to the ability of a search strategy or diagnostic test to identify all relevant cases or studies related to a particular topic or condition. In particular, it measures the number of relevant records identified by a search strategy divided by the total number of relevant records on a given topic.

Sensitivity: Sensitivity is the ability of a diagnostic test to correctly identify individuals who have a particular disease or condition. It is calculated as the proportion of true positive results among all individuals who actually have the condition.

Specificity: Specificity is the ability of a diagnostic test to correctly identify individuals who do not have a particular condition. It is calculated as the proportion of true negative results among all individuals who do not have the condition.

Chapter 4
Beyond the Individual Review: A Practical Guide to Umbrella Review

Mohammed Abdellatif
https://orcid.org/0000-0001-6872-8426
Muscat Private Hospital, Oman

Van Thanh Nguyen
Oxford University Clinical Research Unit, Ho Chi Minh City, Vietnam

Nguyen Hai Nam
https://orcid.org/0000-0001-5184-6936
Cancer Center, Cho Ray Hospital, Ho Chi Minh City, Vietnam

Mohammad Najm Dadam
https://orcid.org/0000-0002-8814-0205
Helios Klinikum Schwelm, Germany

ABSTRACT

Among different research methodologies, the umbrella review offers a very comprehensive tool for synthesizing evidence from multiple systematic reviews with or without meta-analyses. The umbrella review can provide

DOI: 10.4018/979-8-3693-3007-4.ch004

a wider perspective by consolidating findings from various secondary evidence sources and highlighting the current state of evidence within a particular field. This chapter aims to discuss the detailed methodology of the umbrella review, its research applications, benefits, and challenges.

1. FOUNDATIONS OF UMBRELLA REVIEWS

1.1 Evolution of Umbrella Reviews in Research Methodology

The rapid expansion of published systematic reviews in recent years, coupled with the redundancy and overlap in the evidence presented across these reviews, necessitates the need for a novel methodology that can synthesize and summarize their collective findings more efficiently. The increasing need in the medical field to navigate the vast and ever-growing pool of research literature has spurred the emergence of a valuable tool for filtering information overload (Klerings et al., 2015). As a strategic response, umbrella reviews have been developed to effectively address the overwhelming volume of current evidence on specific topics and identify discrepancies in findings (Papatheodorou & Evangelou, 2022).

The methodology known as the "umbrella review" didn't emerge from the contributions of a single individual. Instead, it resulted from collective progress within the scientific community, with significant contributions from figures like John Ioannidis, who played a pivotal role in its development and refinement through various works (Ioannidis, 2009; Ioannidis, 2017). Besides, the collaborative nature of this methodology is exemplified by the efforts of The Joanna Briggs Institute, which established a methodology group to build comprehensive guidelines for conducting umbrella reviews, thereby formalizing the process and ensuring its rigor and reliability (Aromataris et al., 2015). However, many reviews that align with what we now call "umbrella reviews" were conducted even before the term was formally recognized, despite not being explicitly labeled as such (Ioannidis, 2009).

1.2 Definition and Key Principles

An umbrella review is a synthesis of existing multiple systematic reviews with or without meta-analyses covering a topic of research, offering a broader scope than traditional systematic reviews which may focus on specific questions on a specific intervention. Recently, they have become prominent for aggregating data from various reviews to comprehensively address broad research questions. For instance, umbrella reviews can systematically examine multiple interventions about a single topic, offering a comprehensive summary that significantly reduces the time and effort required to read numerous individual systematic reviews (Choi & Kang, 2023).

One of the primary strengths of umbrella reviews lies in their ability to address a wide range of questions, particularly when comprehensive evaluations of various interventions for specific diseases are required or when rapid access to high-quality evidence is essential for developing new policies or guidelines. By synthesizing and summarizing the findings of multiple systematic reviews, umbrella reviews enhance accessibility to relevant evidence and support informed decision-making. Umbrella reviews thus provide valuable guidance for clinicians, healthcare professionals, and policymakers in making evidence-based choices (Pollock et al., 2019).

2. METHODOLOGICAL FRAMEWORK

Figure 1. Detailed flow diagram guideline for umbrella review steps

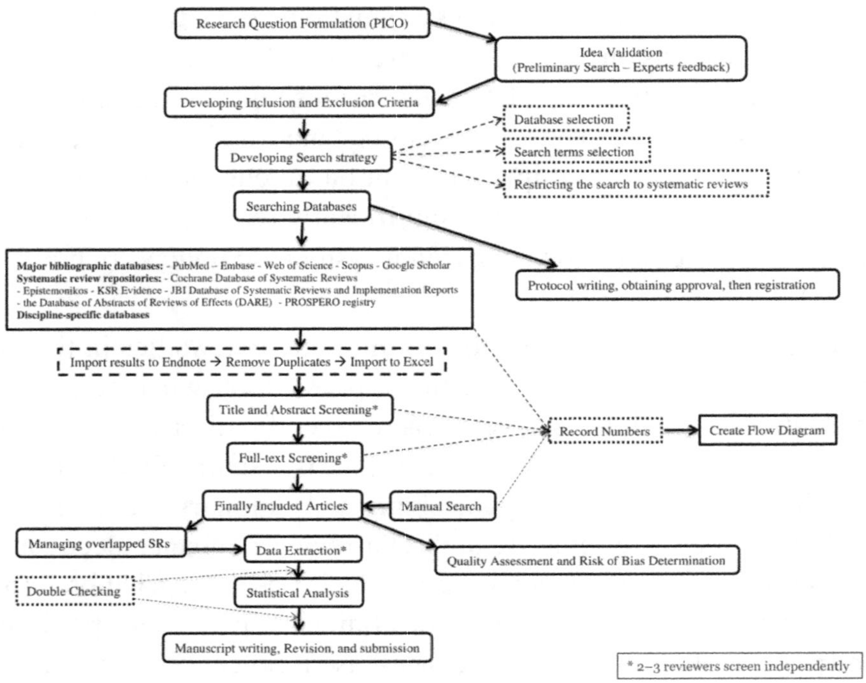

The steps for conducting an umbrella review include the following:

2.1 Developing a Research Question

Building a good quality research depends on developing a well-formulated research question. Start by determining a general subject area, such as a disease, treatment plan, or the target population. Then, reduce the question to its essence using the PICO (Population, Intervention, Comparator, and Outcome) model. After that, you need to make your objectives more specific, for example, by determining whether you are evaluating the safety profiles or effectiveness of different interventions. Then, specify the scope of the review, including study periods, location, and designs, and specify potential subgroups, e.g., different dosages or routes of administration. Then, filter the main elements into a specific question using the previously specified PICO framework and contextual factors.

2.2 Preliminary Research and Idea Validation

At this point, you need to do a preliminary search to ensure that a good number of systematic reviews are available. Afterward, you need to discuss your research question with experts in the field to ensure it is relevant and applicable.

2.3 Developing Inclusion and Exclusion Criteria

The criteria of inclusion and exclusion should be derived from a well-designed PICO framework and contextual factors (Belbasis et al., 2022; Cant et al., 2022; Pollock et al., 2020). The studies included in the umbrella review, are exclusively systematic reviews with or without meta-analyses (Aromataris et al., 2014). It is essential that the definition of a systematic review is clear because not all publications labeled as such meet the defined criteria. Similarly, publications not explicitly labeled as systematic reviews may still meet the criteria (Pollock et al., 2020). Researchers need to determine which type of systematic reviews to include. This decision involves choosing between exclusively incorporating systematic reviews of randomized controlled trials (RCTs) or including other study designs, such as observational studies. The choice should align with the research question and the available evidence (Pollock et al., 2020). Clarifying inclusion and exclusion criteria will help the reader understand the purpose and context of the umbrella review. It will also guide the authors themselves in the study's selection step of the umbrella review (Aromataris et al., 2014).

2.3.1 Participants

Detailed characteristics of the study participants should be outlined, this might include age (e.g., neonates, pediatrics, geriatrics, etc.,), disease severity, *in vitro* or animal studies, and other identifying characteristics. The intention of doing a separate analysis of a certain subpopulation should be specified in advance and the subpopulation should be adequately defined as well. In addition, a definition of the disease or the condition under investigation should be described as well as the setting in which the study is conducted (e.g., hospital, acute care, community, etc.,). Both inclusion and exclusion participants' characteristics should be justified.

2.3.2 Interventions/Comparator

In interventional studies, the intervention(s) of interest as well as the comparator(s) used in the studies should be defined in detail. It's essential to ensure both experimental and control (comparator) arms have been sufficiently defined with consideration of variations within the interventions (e.g., dosage, duration, route of administration, etc.,). The control arm might be inactive (e.g., placebo, no intervention, routine care), or active interventions (e.g., different dose or route of administration, another intervention, etc.). It is important to take into account the different interpretations of terms like ‘control’, ‘placebo’, ‘no intervention’, or ‘usual care’.

Authors need to decide whether they wish to include studies where the intervention of interest is combined with another intervention or not.

2.3.3 Outcomes

It is important to define beforehand the outcomes that will be measured and reported. These outcomes need to directly relate to the review's main question and be significant for the group being studied. If there are surrogate outcomes used, they should be clarified and shown to be linked to outcomes that matter to patients (Aromataris et al., 2014).

Outcomes reported shouldn't often decide if a study is included in a review. Specifically, studies shouldn't be left out just because they didn't share results for a measured outcome or offer 'unusable data'. This is important to prevent bias that results from authors selectively reporting their findings (Aromataris et al., 2020).

To ensure a comprehensive understanding of a topic and support decision-making, the umbrella review should cover both positive and negative outcomes reported in the evidence (Aromataris et al., 2014).

2.4 Database Search for Systematic Reviews

Database search is a crucial step in conducting systematic reviews. It involves careful selection of databases, choosing appropriate search terms, and restricting the search to focus on systematic reviews.

2.4.1 Database Selection (Hartling et al., 2016)

There are major bibliographic databases that do not include systematic reviews specifically. These include PubMed, EMBASE, Scopus, Web of Science, and others. However, there are some systematic review repositories that must be searched. These are listed as follows with their respective links:

a) *Cochrane Database of Systematic Reviews (*https://www.cochranelibrary.com/*)*
b) Epistemonikos (https://www.epistemonikos.org/)
c) KSR Evidence (https://ksrevidence.com/)
d) JBI Database of Systematic Reviews and Implementation Reports (https://library.sfsu.edu/joanna-briggs-institute-database-systematic-reviews-implementation-reports-2012-2019)
e) DARE (https://www.crd.york.ac.uk/CRDWeb/)
f) PROSPERO register (https://www.crd.york.ac.uk/prospero/)

In addition, there are some discipline-specific databases that can be searched according to the research question (Pollock et al., 2020). Examples of these databases are listed below:

a) PsycINFO (American Psychological Association)
b) LILACS (scientific health information from Latin America and the Caribbean countries)
c) CINAHL (Nursing and Allied Health Literature)
d) PEDro (physiotherapy reviews)
e) OTseeker (Occupational Therapy reviews)
f) BEME and the EPPI Centre Evidence Library (education-related reviews)

- **Grey literature**: For a more comprehensive umbrella review, it is important to seek out grey literature or unpublished reports. Many of these reports are available on government or organizational websites (Aromataris et al., 2014).

2.4.2 Selecting Search Terms

Researchers need to carefully choose relevant search terms, including controlled vocabulary terms (e.g., MeSH) and free-text terms (Belbasis et al., 2022; Pollock et al., 2020). The search terms should accurately reflect the research question formulated earlier using the PICO framework and the concepts being studied. The initial search terms should be used for a trial search, then the authors can look for other keywords within the relevant retrieved systematic reviews. It might take some rounds of search and refinement of search terms, till the final search terms are formulated.

2.4.3 Restricting the Search to Systematic Reviews

The aim is to optimize the search strategy by minimizing the inclusion of non-systematic review publications (Aromataris et al., 2015; Cooper & Koenka, 2012; Pollock et al., 2016; Smith et al., 2011). This can be achieved by using specific search filters or limiters, such as the "Systematic Reviews" filter in PubMed or utilizing MeSH headings and search terms specifically targeting systematic reviews, such as "systematic review" and "meta-analysis".

It's important to underscore that while utilizing filters, MeSH terms, and subheadings can enhance the precision of literature searches, there's a potential drawback: the categorization process within databases may be time-consuming, often spanning up to nine months. Consequently, this delay poses a risk of overlooking pertinent articles, especially those recently published or not yet classified. Hence, emphasizing the significance of keyword-based searches remains imperative in developing comprehensive search strategies.

In response to these challenges, some authors developed custom search filters tailored for Pubmed. These filters led to a notable 61.3% surge in the number of articles retrieved, which could potentially be systematic reviews. Additionally, they attained an average precision likely exceeding 80% (Salvador-Oliván et al., 2021).

2.5 Selecting Eligible Systematic Reviews

Deciding which systematic reviews to incorporate into the umbrella review holds significant influence over the review process. To ensure accuracy, at least two authors should work independently to assess whether each study aligns with the eligibility criteria (Edwards et al., 2002; Gartlehner et al., 2020; Waffenschmidt et al., 2019).

Selecting eligible systematic reviews involves the following steps:

(i) Combining search results from different databases into one list

This can be done using a reference management software program, such as EndNote or Mendeley. It is important to remove any duplicate records from the list.

(ii) Screening titles and abstracts

This step involves a thorough review of titles and abstracts of retrieved records. Studies that are clearly irrelevant should be excluded at this stage. In this phase, erring on the side of inclusivity might be advantageous.

(iii) Downloading the full text of the potentially relevant reports
(iv) Screening the full texts against eligibility criteria

It is recommended to train the screening team by conducting a pilot screening task in which the authors screen a small number of studies against the eligibility criteria. This minimizes errors resulting from misunderstanding of the inclusion/exclusion criteria. Remember that failure of one inclusion criterion leads to exclusion of the review and if a potentially relevant systematic review has a broader research question than the umbrella review, evaluate the primary studies included in this review against the umbrella review's inclusion criteria.

(v) Resolving possible conflicts

Disagreements, which usually result from minor oversight from one of the authors conducting the review, are usually resolved through discussion. However, if the disagreement results from differing interpretations, it might necessitate a discussion with a senior author for resolution (Pollock et al., 2020).

(vi) Creating a flow diagram

This requires detailed documentation of the selection process. It's essential to maintain a record of the number of references and studies throughout the selection process by using a simple spreadsheet, more complex reference management tools, or specialized systematic review software

Figure 2. Flow diagram of systematic reviews identification and selection for the umbrella review

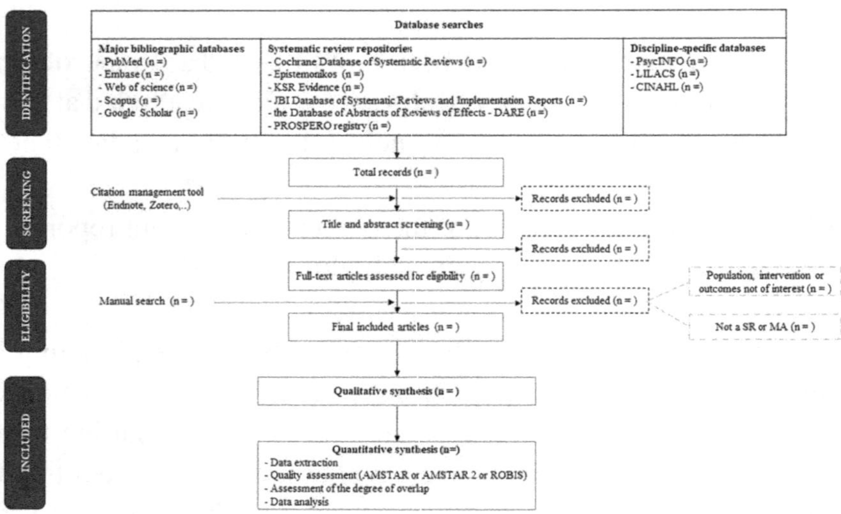

2.5.1 Manual Search

Following the identification of included studies, manual search methods play an important role in supplementing the gathered information. These methods include: exploring the references cited within the included reviews, contacting authors of the included studies, experts in the field, or manufacturers of related products, looking at the related articles/searches in PubMed and Google Scholar, and looking at papers that cite the paper on Google Scholar.

2.6 Data Collection, Analysis, and Presentation from Included Systematic Reviews

Once the relevant systematic reviews have been identified, data collection, analysis, and presentation can be carried out.

2.6.1 Data Collection

Initially, a pilot extraction sheet should be generated using random reviews. If necessary, subsequent modifications should be discussed among all reviewers (Aromataris et al., 2014; Pollock et al., 2020). Two to three researchers should independently extract the required data using a standardized data extraction form. In the event of any disagreements, they should be resolved through discussion and consensus (Belbasis et al., 2022; Pollock et al., 2020).

Data extraction should encompass information regarding systematic reviews and the primary studies they comprise. Generally, these data can be categorized into four main groups: basic characteristics, quantitative outcomes, data for evaluating the risk of bias, and data for assessing the certainty of evidence (Ballard & Montgomery, 2017; Pollock et al., 2016). Table 1 contains data that are typically extracted from the included articles.

Table 1. Data typically included in the data extraction form

Data in the extraction form:
- Authors' names - Publication year - The number of databases searched - Search period - Review objectives (primary and secondary outcomes) - Interventions and control arms (type, dose, intensity, frequency, and duration) - Results synthesis method - Key findings - Types of studies contributing to outcomes and their countries of origin - Quality assessment instrument and ratings of each systematic review as well as the included primary studies. - Basic information about the included primary studies

Authors should plan ahead for missing or inadequately reported data (Ballard & Montgomery, 2017; Pollock et al., 2016). They can either acknowledge data gaps in their umbrella review or extract missing data from primary studies. Extracting a significant amount of data from primary studies may prompt authors to consider a systematic review instead of an umbrella review (Pollock et al., 2020). For systematic reviews without meta-analyses, the rationale for not conducting meta-analysis should be stated (Belbasis et al., 2022).

2.6.2 Data Checking

After the data extraction sheet is ready and before proceeding with the analysis, it is important to double-check the data between the original article and corresponding fields in the sheet to ensure the accuracy of the data and to avoid any possible human error. This should be typically conducted by 2-3 independent reviewers and ideally, each author should check systematic reviews different from the ones he previously extracted.

2.6.3 Data Analysis, and Presentation

There are two main ways to present outcome data in an umbrella review: "*narrative summary*" which provides a descriptive overview and "*repeated analysis*" which allows for additional analyses and comparisons (Pollock et al., 2020). Choose the appropriate method based on purpose, topic area, and characteristics of included systematic reviews. The differences between the two approaches are highlighted in Table 2. Finally, present a clear and concise summary of the results, utilizing tables, figures, or graphs to present the synthesized data and highlight key findings.

Table 2. Key differences between the two methods of data presentation in the umbrella reviews

Approach	Description	Purpose	Methodological considerations
Narrative summary	Presenting data exactly as reported in included SRs, including narrative summaries and meta-analyzed data	Provide an overview of SR content and results, describe interventions, comparators, outcomes, and results	Extract and reformat effect estimates, 95% confidence intervals, and measures of heterogeneity from SR
Repeated analysis	Extracting relevant data from SRs and conducting different analyses than original reviews	Answer different clinical questions, apply consistent analyses, address differences in summary measures or models, analyze data not previously meta-analyzed	Adhere to standard meta-analytic principles, exercise caution to avoid inappropriate comparisons, understand reasons behind original review authors' analytic methods

SR: Systematic Review

2.7 Quality Assessment and Risk of Bias of Included Systematic Reviews

In conducting an umbrella review, it is an important consideration to evaluate the methodological quality and risk of bias of the included systematic reviews. This step ensures the credibility and reliability of the synthesized evidence. At least two reviewers should independently conduct quality assessments.

Researchers can utilize established tools such as AMSTAR, AMSTAR2, and ROBIS to achieve this.

2.7.1 AMSTAR (A MeaSurement Tool to Assess Systematic Reviews)

AMSTAR was developed in 2007 and is designed to appraise the methodological quality of systematic reviews focusing on randomized controlled trials (RCTs). It consists of 11 items that cover various aspects of the review process, including study selection, data extraction, and consideration of publication bias. AMSTAR has been proven to be a reliable and valid tool for evaluating the quality of systematic reviews (Shea et al., 2007).

2.7.2 AMSTAR2

An updated version of AMSTAR, AMSTAR2 expands its scope to assess the methodological quality of systematic reviews that include both randomized and non-randomized studies of healthcare interventions. It contains 16 domains in total, including 10 from the original tool. This tool includes items such as protocol registration, adequacy of literature search, risk of bias assessment, and appropriateness of meta-analytic methods. Additionally, AMSTAR2 provides criteria for rating the overall confidence in the study, enabling a more comprehensive evaluation (Shea et al., 2017).

2.7.3 ROBIS (Risk of Bias Assessment Tool for Systematic Reviews)

ROBIS was developed in 2016 to assess the risk of bias in systematic reviews. It addresses questions related to interventions, diagnosis, prognosis, and etiology studied in the reviews. ROBIS consists of three phases: assessing relevance, identifying concerns with the review process, and judging the overall risk of bias. This tool offers a valuable framework for evaluating potential biases in systematic reviews (Whiting et al., 2016).

Researchers should select the appropriate tool based on their specific research design and preferences (Pollock et al., 2020). While some studies have shown that AMSTAR is frequently used in umbrella reviews (Pieper et al., 2012; Pollock et al., 2016), there is currently no consensus on the most effective tool. Therefore, authors should exercise their judgment in choosing the most suitable tool for their particular review and research objectives.

2.8 Managing Overlapping Systematic Reviews

The identification and management of overlapping systematic reviews is a very important step in the synthesis of umbrella reviews (Siontis et al., 2013). Systematic reviews are considered overlapping if they share some or all of the same primary studies. The inclusion of these studies multiple times in the final umbrella review will lead to an exaggeration of the final calculated effect size (Lunny et al., 2021). A systematic approach must be planned to carefully search for these reviews among the included articles. This can be done using a matrix or a spreadsheet in which each column represents a systematic review and each row represents a primary study included in this review (Hennessy & Johnson, 2020; Lunny et al., 2017).

The management of overlapping reviews can be done by several approaches (Gutierrez-Arias et al., 2023; Pollock et al., 2020). One option is to select one of the overlapping reviews and discard the others. The selected review might be different for each of the umbrella review questions. In this case, the authors may choose the most comprehensive review in answering this question, the most recent one or the one with the highest quality. Another option is to represent the findings of each review in a narrative way and try to discuss and possibly explain the possible discrepancies between the reviews' findings (De Caro et al., 2016)

2.9 Manuscript Writing

This usually follows the same steps of manuscript writing explained in a separate chapter of this book

3. BENEFITS AND CHALLENGES

3.1 Advantages of Employing Umbrella Reviews

The recent increase in the number of primary and secondary evidence sources challenges readers to source, collate, and interpret the evidence, which sometimes reveals conflicting results (Cant et al., 2022). Umbrella reviews systematically synthesize evidence from multiple systematic reviews with or without meta-analyses on a specific research topic (Higgins et al., 2023). Among several levels of scientific evidence, Umbrella reviews are at the highest levels currently available and provide potential benefits if performed and appropriately interpreted (Belbasis et al., 2022). Firstly, by using the existing systematic review and meta-analysis, umbrella reviews can save the research resources that would be wasted on re-conducting and scraping from the primary studies. From this, they also help to avoid duplication, reduce repetition, and minimize the waste of research efforts from overlapped research ideas and the exponential of conducting systematic reviews and meta-analyses. Next, umbrella reviews highlight the knowledge gaps in a specific research area or the uncertainty, inconsistency, or even contrast in the literature and can provide suggestions for future research. Furthermore, the umbrella reviews present an overview of the included systematic reviews' quality, effect sizes, uncertainty, and various further outcomes. They can also account for heterogeneity, bias indicators, and confounding factors that might not be addressed or neglected.

Figure 3. Hierarchy of scientific research design

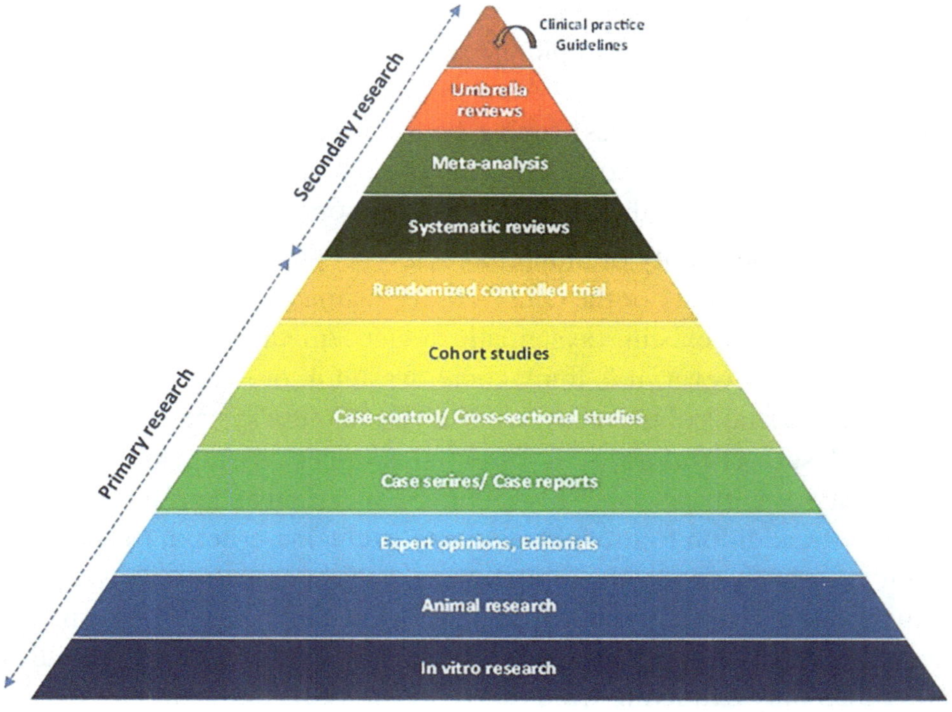

Umbrella reviews can present a comprehensive and integrated overview of the evidence on a research topic by applying standardized methods to select, appraise, and re-analyze the available systematic reviews/meta-analyses. Consequently, they present and compare evidence, offering a comprehensive view of the evidence's strengths and limitations.

3.2 Common Challenges and Potential Solutions

Umbrella reviews are considered to provide the highest level of evidence for scientific research, but they also encounter several challenges.

One of the most important challenges is the dependency of umbrella reviews on the number of existing evidence base, which comprises studies that have been published and integrated into systematic reviews with or without meta-analyses (Gates et al., 2020). There may be unpublished or under-reviewed systematic reviews/meta-analyses or primary studies that the umbrella review did not capture at the time of their conduct. To address this issue, an umbrella review could include all relevant studies on a well-defined research topic, irrespective of their inclusion in prior reviews (Sanders et al., 2023). Another challenge is the currency of the evidence base, which requires keeping the search for publications as updated as possible. This implies conducting a thorough literature search for each subdomain included in the umbrella review (López-López et al., 2022).

Moreover, the quality of umbrella reviews is primarily determined by the quality and validity of the data from the primary studies that are used to synthesize the existing systematic reviews/meta-analyses (Belbasis et al., 2022). However, it is impractical, wasteful, or even impossible to verify all the details of the analyzed database in these systematic reviews/ meta-analyses. Therefore, it is essential to select systematic reviews/ meta-analyses with caution. This should be specified and proposed in the protocol. In addition to the inclusion and exclusion criteria, the authors should rigorously review the limitations of each systematic review- meta-analysis and a consensus should be reached at each stage. Furthermore, the authors should randomly check information in the original studies rather than only extracting from the systematic reviews/meta-analyses.

Another challenge is the heterogeneity of eligibility criteria across different levels. The selection of primary studies for systematic review/ meta-analyses may vary depending on the aim of each systematic review/ meta-analyses. These systematic reviews/meta-analyses are then assessed for inclusion in the umbrella reviews using various eligibility criteria, which introduces another level of heterogeneity. These deviations may result from the study population, study design (e.g., randomized trials, cohort studies, case-control studies, etc.), interventions, or the primary objectives that have been overlooked in the systematic review/ meta-analyses (Belbasis et al., 2022). To address this issue, an analysis plan should be specified and proposed before conducting the umbrella review.

Finally, the unit of analysis for the umbrella review is the systematic reviews and/or meta-analyses of the included studies; hence, the umbrella review will inherit the limitations of these systematic reviews/meta-analyses. Therefore, caution is warranted when interpreting the findings. For instance, if the eligible systematic review/meta-analyses examined the association of a disease or phenomenon, it would be erroneous for the umbrella review to infer the causation of this disease or phenomenon (Fusar-Poli & Radua, 2018). Furthermore, the primary studies may have various sources of bias and confounding that are difficult to detect or adjust for in systematic review meta-analyses, and these may persist in the umbrella review. The authors should be mindful of these issues when interpreting and discussing the results, as well as the implications for generalizability(Belbasis et al., 2022; Choi & Kang, 2023).

4. CONCLUSION

As a result of the rapid expansion in research publications as well as increasing numbers of systematic reviews, umbrella review have been developed to provide a comprehensive understanding of specific topics. However, their value depends on accurate execution, prompting this guide for researchers in conducting umbrella reviews.

REFERENCES

Aromataris, E., Fernandez, R., Godfrey, C., Holly, C., Khalil, H., & Bhatarasakoon, P. (2020). Umbrella reviews. In Aromataris, E., & Munn, Z. (Eds.), *JBI manual for evidence synthesis*. JBI.

Aromataris, E., Fernandez, R., Godfrey, C. M., Holly, C., Khalil, H., & Tungpunkom, P. (2015). Summarizing systematic reviews: Methodological development, conduct and reporting of an umbrella review approach. *JBI Evidence Implementation*, *13*(3), 132–140. DOI: 10.1097/XEB.0000000000000055 PMID: 26360830

Aromataris, E., Fernandez, R. S., Godfrey, C., Holly, C., Khalil, H., & Tungpunkom, P. (2014). *Methodology for JBI umbrella reviews*. Joanna Briggs Institute.

Ballard, M., & Montgomery, P. (2017). Risk of bias in overviews of reviews. *Research Synthesis Methods*, *8*(1), 92–108. DOI: 10.1002/jrsm.1229 PMID: 28074553

Belbasis, L., Bellou, V., & Ioannidis, J. P. A. (2022). Conducting umbrella reviews. *BMJ Medicine*, *1*(1), e000071. DOI: 10.1136/bmjmed-2021-000071 PMID: 36936579

Cant, R., Ryan, C., & Kelly, M. A. (2022). A nine-step pathway to conduct an umbrella review of literature. *Nurse Education in Practice*, *32*(2), 31–34. DOI: 10.1111/nae2.12039

Choi, G. J., & Kang, H. (2023). Introduction to umbrella reviews as a useful evidence-based practice. *Journal of Lipid and Atherosclerosis*, *12*(1), 3–11. DOI: 10.12997/jla.2023.12.1.3 PMID: 36761061

Cooper, H., & Koenka, A. C. (2012). The overview of reviews: Unique challenges and opportunities when research syntheses are the principal elements of new integrative scholarship. *The American Psychologist*, *67*(6), 446–462. DOI: 10.1037/a0027119 PMID: 22352742

De Caro, W., Marucci, A. R., Lancia, L., & Sansoni, J. (2016). Umbrella reviews. In *Evidence synthesis with overviews of reviews and meta-epidemiologic studies* (pp. 150–180). Springer.

Edwards, P., Clarke, M., DiGuiseppi, C., Pratap, S., Roberts, I., & Wentz, R. (2002). Identification of randomized controlled trials in systematic reviews: Accuracy and reliability of screening records. *Statistics in Medicine*, *21*(11), 1635–1640. DOI: 10.1002/sim.1190 PMID: 12111924

Fusar-Poli, P., & Radua, J. (2018). Ten simple rules for conducting umbrella reviews. *Evidence-Based Mental Health*, *21*(3), 95–100. DOI: 10.1136/ebmental-2018-300014 PMID: 30006442

Gartlehner, G., Affengruber, L., Titscher, V., Noel-Storr, A., Dooley, G., Ballarini, N., & König, F. (2020). Single-reviewer abstract screening missed 13 percent of relevant studies. *Journal of Clinical Epidemiology*, *121*, 20–28. DOI: 10.1016/j.jclinepi.2020.01.005 PMID: 31972274

Gates, M., Gates, A., Guitard, S., Pollock, M., & Hartling, L. (2020). Guidance for overviews of reviews continues to accumulate. *Systematic Reviews*, *9*(1), 254. Advance online publication. DOI: 10.1186/s13643-020-01509-0 PMID: 33148319

Gutierrez-Arias, R., Pieper, D., Lunny, C., Torres-Castro, R., Aguilera-Eguía, R., & Seron, P. (2023). Strategies used to manage overlap of primary study data by exercise-related overviews: Protocol for a systematic methodological review. *BMJ Open*, *13*(4), e069906. DOI: 10.1136/bmjopen-2022-069906 PMID: 37080626

Hartling, L., Featherstone, R., Nuspl, M., Shave, K., Dryden, D. M., & Vandermeer, B. (2016). The contribution of databases to the results of systematic reviews: A cross-sectional study. *BMC Medical Research Methodology*, *16*(1), 127. Advance online publication. DOI: 10.1186/s12874-016-0232-1 PMID: 27670136

Hennessy, E. A., & Johnson, B. T. (2020). Examining overlap of included studies in meta-reviews. *Research Synthesis Methods*, *11*(1), 134–145. DOI: 10.1002/jrsm.1390 PMID: 31823513

Ioannidis, J. P. A. (2009). Integration of evidence from multiple meta-analyses: A primer on umbrella reviews, treatment networks and multiple treatments meta-analyses. *Canadian Medical Association Journal*, *181*(8), 488–493. DOI: 10.1503/cmaj.081086 PMID: 19654195

Ioannidis, J. P. A. (2017). Next-generation systematic reviews: Prospective meta-analysis, individual-level data, networks and umbrella reviews. *British Journal of Sports Medicine*, *51*(20), 1456–1458. DOI: 10.1136/bjsports-2017-097621 PMID: 28223307

Klerings, I., Weinhandl, A. S., & Thaler, K. J. (2015). Information overload in healthcare: Too much of a good thing? *Zeitschrift für Evidenz, Fortbildung und Qualität im Gesundheitswesen*, *109*(4–5), 285–290. DOI: 10.1016/j.zefq.2015.06.005 PMID: 26354128

López-López, J. A., Rubio-Aparicio, M., & Sánchez-Meca, J. (2022). Overviews of reviews: Concept and development. *Psicothema*, *34*(2), 175–181. DOI: 10.7334/psicothema2021.586 PMID: 35485529

Lunny, C., Brennan, S. E., McDonald, S., & McKenzie, J. E. (2017). Toward a comprehensive evidence map of overview methods. *Systematic Reviews*, *6*, 231. Advance online publication. DOI: 10.1186/s13643-017-0617-1 PMID: 29162130

Lunny, C., Pieper, D., Thabet, P., & Kanji, S. (2021). Managing overlap of primary study results across systematic reviews. *BMC Medical Research Methodology*, *21*(1), 140. Advance online publication. DOI: 10.1186/s12874-021-01269-y PMID: 34233615

Papatheodorou, S. I., & Evangelou, E. (2022). Umbrella reviews: What they are and why we need them. In *Methods in molecular biology* (Vol. 2345, pp. 135–146). Springer., DOI: 10.1007/978-1-0716-1566-9_8

Pieper, D., Buechter, R., Jerinic, P., & Eikermann, M. (2012). Overviews of reviews often have limited rigor. *Journal of Clinical Epidemiology*, *65*(12), 1267–1273. DOI: 10.1016/j.jclinepi.2012.06.015 PMID: 22959594

Pollock, M., Fernandes, R. M., Becker, L. A., Featherstone, R., & Hartling, L. (2016). What guidance is available for researchers conducting overviews of reviews of healthcare interventions? *Systematic Reviews*, *5*(1), 190. Advance online publication. DOI: 10.1186/s13643-016-0367-5 PMID: 27842604

Pollock, M., Fernandes, R. M., Becker, L. A., Pieper, D., & Hartling, L. (2020). Overviews of reviews. In *Cochrane handbook for systematic reviews of interventions* (Version 6). Cochrane.

Pollock, M., Fernandes, R. M., Newton, A. S., Scott, S. D., & Hartling, L. (2019). A decision tool to help researchers make decisions about including systematic reviews in overviews of reviews of healthcare interventions. *Systematic Reviews*, *8*(1), 29. Advance online publication. DOI: 10.1186/s13643-018-0768-8 PMID: 30670086

Salvador-Oliván, J. A., Marco-Cuenca, G., & Arquero-Avilés, R. (2021). Development of an efficient search filter to retrieve systematic reviews from PubMed. *Journal of the Medical Library Association: JMLA*, *109*(4), 561–570. DOI: 10.5195/jmla.2021.1223 PMID: 34858085

Sanders, T., Noetel, M., Parker, P., Del Pozo Cruz, B., Biddle, S., Ronto, R., Hulteen, R., Parker, R., Thomas, G., De Cocker, K., Salmon, J., Hesketh, K., Weeks, N., Arnott, H., Devine, E., Vasconcellos, R., Pagano, R., Sherson, J., Conigrave, J., & Lonsdale, C. (2023). An umbrella review of the benefits and risks associated with youths' interactions with electronic screens. *Nature Human Behaviour*, *8*(1), 82–99. Advance online publication. DOI: 10.1038/s41562-023-01712-8 PMID: 37957284

Shea, B. J., Grimshaw, J. M., Wells, G. A., Boers, M., Andersson, N., Hamel, C., Porter, A. C., Tugwell, P., Moher, D., & Bouter, L. M. (2007). Development of AMSTAR: A measurement tool to assess the methodological quality of systematic reviews. *BMC Medical Research Methodology*, *7*(1), 10. Advance online publication. DOI: 10.1186/1471-2288-7-10 PMID: 17302989

Shea, B. J., Reeves, B. C., Wells, G., Thuku, M., Hamel, C., Moran, J., Moher, D., Tugwell, P., Welch, V., Kristjansson, E., & Henry, D. A. (2017). AMSTAR 2: A critical appraisal tool for systematic reviews. *BMJ (Clinical Research Ed.)*, *358*, j4008. DOI: 10.1136/bmj.j4008 PMID: 28935701

Siontis, K. C., Hernandez-Boussard, T., & Ioannidis, J. P. A. (2013). Overlapping meta-analyses on the same topic. *BMJ (Clinical Research Ed.)*, *347*(jul19 1), f4501. DOI: 10.1136/bmj.f4501 PMID: 23873947

Smith, V., Devane, D., Begley, C. M., & Clarke, M. (2011). Methodology in conducting a systematic review of systematic reviews of healthcare interventions. *BMC Medical Research Methodology*, *11*(1), 15. Advance online publication. DOI: 10.1186/1471-2288-11-15 PMID: 21291558

Waffenschmidt, S., Knelangen, M., Sieben, W., Bühn, S., & Pieper, D. (2019). Single screening versus conventional double screening for study selection in systematic reviews. *BMC Medical Research Methodology*, *19*(1), 132. Advance online publication. DOI: 10.1186/s12874-019-0782-0 PMID: 31253092

Whiting, P., Savović, J., Higgins, J. P. T., Caldwell, D. M., Reeves, B. C., Shea, B., Davies, P., Kleijnen, J., & Churchill, R. (2016). ROBIS: A new tool to assess risk of bias in systematic reviews. *Journal of Clinical Epidemiology*, *69*, 225–234. DOI: 10.1016/j.jclinepi.2015.06.005 PMID: 26092286

Chapter 5
Identifying Scope and Research Question

Federica Cuce
https://orcid.org/0000-0002-7443-1976
University of Verona, Italy

Celine Tabche
https://orcid.org/0000-0001-7690-6368
Imperial College London, UK

Nguyen Quan
UCL Institute for Global Health, UK

ABSTRACT

Scope and research questions must be the guiding light for the definition of the research methodology and the development of a review project. A well-defined scope ensures that research objectives remain focused and attainable by considering existing knowledge, its gaps, and potential novel contributions to the field. A rigorous formulation of a structured research question should follow a successful scope definition aiming at a meaningful contribution to the chosen field. While maintain the specificity of the research focus it's important to allow adjustments based on unforeseen challenges and emerging trends in research and policymaking, integrating them into the scope and research question while maintaining

DOI: 10.4018/979-8-3693-3007-4.ch005

relevance and impact. To achieve this, stakeholder engagement should be integrated in research project planning. In summary, a meticulously crafted scope and research question are pillars of a review, fundamental to provide reliable information to research users applicable in a real setting to inform decision-making.

INTRODUCTION

To start a research project, researchers must understand current gaps in the literature and how to effectively fill them and solve end users' questions and problems. When defining the outline of a systematic review, one must start with a complete definition of its focus and clear and comprehensive questions to avoid ambiguity among researchers, stakeholders, and funders.

Developing a well-structured research question for a review can involve multiple discussions and should be achieved following stringent criteria for clarity and completeness. Stakeholders' needs and opinions are necessary to achieve relevant research that can impact clinical practice.

Review questions are the basis for identifying the type of studies to review and information to extract, structure data synthesis, and disseminate results. They should be aligned with the subsequent choices of methodology. They should be able to maintain a degree of flexibility to be feasible while maintaining the original framework on scope and research priorities to remain relevant to the research panorama gaps and for end-users.

This chapter will explore the nuances of well-formulated questions and how focusing on this step is strategic to guide the following steps of a review. It will provide practical insights through techniques and tools that are easy to apply and remember, specific advice, and strategies to ally in the pivotal initial stages of a research project.

THE IMPORTANCE OF THE SCOPE

Research scope is a fundamental concept in the research process, defining the boundaries and parameters within which a study operates. It encompasses the breadth and depth of the investigation, guiding researchers in their pursuit of answering specific research questions and achieving predetermined objectives (Grant & Booth, 2009). The definition of research scope is multifaceted, encapsulating the objectives, target population, geographic location, time frame, and methodology of the study (Bhattacherjee, 2012). Essentially, it serves as a roadmap, delineating the extent of inquiry and providing a framework for decision-making throughout the research journey.

The role of scope in delineating research boundaries cannot be overstated. By clearly defining the scope, researchers establish the parameters within which their study operates, thus preventing scope creep and maintaining research focus (Creswell & Creswell, 2017). Scope delineation helps researchers identify what is within their investigation and what falls outside, facilitating effective resource allocation and time management. Moreover, it enables researchers to manage expectations, both within the research team and among stakeholders, by clearly articulating the intended scope of the study (Hart, 1998).

The impact of a well-defined scope on research clarity and focus is profound. A clear scope ensures that research objectives remain explicit and attainable, giving researchers a sense of direction and purpose (Yin, 2017). It fosters clarity of thought and precision in execution, minimizing ambiguity and confusion throughout the research process (Neuman, 2014). Furthermore, a well-defined scope enhances research focus by guiding researchers in selecting relevant literature, data collection methods, and analytical techniques (Miles et al., 2014). As a result, researchers are better equipped to generate meaningful insights and draw valid conclusions from their findings.

A clear definition of the scope should be identified in the first steps of the review ideation and referenced during the definition of the subsequent roadmap: research question, inclusion criteria and all the steps of the review process. It can start from an existing framework of ideas, discussed research priorities, or advice on further research in other works. Although short in nature, its wording should be carefully chosen by the research team and shared with funders and stakeholders before the beginning of the project to avoid redirections in the process that might be secondary to a less thorough analysis of the problem, the priorities or the feasibility of the project.

In summary, understanding the importance of scope is paramount in conducting rigorous and impactful research. By defining the scope, researchers establish the boundaries and parameters within their study, thereby preventing scope creep and maintaining research focus. A well-defined scope enhances research clarity and focus, providing researchers with a roadmap for decision-making and guiding them in answering specific research questions.

NAVIGATING THE RESEARCH LANDSCAPE

Navigating the research landscape involves a comprehensive assessment of existing knowledge, identifying gaps, and exploring potential contributions to the field. This process is essential for defining the scope of a research study and ensuring its relevance and significance (Petticrew & Roberts, 2008). By considering the existing literature, researchers can build upon previous findings, identify areas needing further investigation, and contribute new insights to the body of knowledge (Greenhalgh, 2014).

Notably, the research landscape is dynamic, continually evolving in response to emerging trends, technological advancements, and societal changes. As such, defining the scope of a study requires careful consideration of current developments and future projections, as well as an awareness of potential shifts in research priorities (Mayer, 2018). Failure to adapt to the changing landscape may result in outdated or irrelevant research findings, undermining the study's impact and applicability (Barbour, 2001).

The importance of context in defining the research scope must be considered. Contextual factors, such as cultural norms, socioeconomic conditions, and political dynamics, shape the research landscape and influence the interpretation and applicability of research findings (Smith, 2015). By acknowledging and accounting for contextual nuances, researchers can ensure the relevance and generalizability of their findings, thereby maximizing the impact of their research.

In navigating the research landscape, researchers must remain vigilant to potential biases and limitations inherent in the existing literature (Gough et al., 2017). By critically evaluating the quality and reliability of sources, researchers can identify gaps in knowledge and areas for further exploration, thus informing the development of research questions and objectives (Booth et al., 2016). Additionally, researchers should seek diverse perspectives and interdisciplinary insights to enrich their understanding of complex phenomena and foster innovative research approaches (Patton, 2015).

To achieve a successful analysis of the current research panorama, it can be useful to rely on updated reviews, meta-analysis, and commentaries, or to conduct a rapid review focusing on the most meaningful papers available, especially if they advise future researchers on current knowledge gaps. Stakeholders' involvement on the other end could help understand the end user perspective on lacking areas, might be more meaningful and obvious to patients or clinicians compared to researchers.

In summary, navigating the research landscape is a multifaceted process that requires careful consideration of existing knowledge, gaps, and potential contributions to the field. By embracing the dynamic nature of the research landscape and contextual factors that shape it, researchers can define the scope of their study in a manner that is relevant, impactful, and responsive to emerging trends and evolving priorities.

DEVELOPING A RESEARCH QUESTION

A clinical question, a gap in the literature or a plethora of different information can prompt researchers to start thinking about a review that can aid themselves and future clinicians to understand the currently available data comprehensively and more clearly. That initial idea has to be framed to become a clear clinical question that can be answered through a comprehensive analysis of the published data. A good research question should be developed with a comprehensive view of the problem (to the greatest possible extent) through a preliminary analysis, using imagination and remaining skeptical.

Before the creation of the definitive formulation of the question(s) researches should understand which stakeholders (patients, health care providers, governments, etc.) are affected by the problem and care about the results of the project. With the involvement of stakeholders in the decision-making process they should find which questions are most important to answer, considering known advantages and disadvantages of the analyzed intervention(s), and how to refine them to develop a feasible research outline. From an understanding of the gaps to focus on, involved parties should agree on the breadth of the review, possible subset of questions or analyses and how to account for changes in focus as the review progresses; and considerations about how broad (or narrow) a review should be.

Crafting a research question that is specific and well-defined to the problem researchers are tackling in the research is paramount to advancing to other steps of the project. From a good research question stems an outline of what the search terms should focus on, the inclusion and exclusion criteria, the data that should be extracted and a structure of the possible analyses and syntheses of findings (Cooper, 1984, Hedges, 1994, Oliver et al., 2017). Moreover, healthcare workers and other researchers that read the review question can understand right from the beginning if the review is relevant to their clinical question.

The PICO mnemonic is often used to consider all the parts necessary to a complete research question. It refers to Population, Intervention, Comparison(s) and Outcome. All the components should be present, but they can be defined more broadly or strictly depending on the voice of focus of the review. They can focus on different areas of patient care. Usually, they are distinguished in Diagnosis, Therapy, Prognosis, and Harm/Etiology. (Richardson et al 1995, Counsell 1997)

To achieve a clear indication of the primary and secondary objectives they should be described indicating all these parts. For example, one could present a study aim focused on therapeutic outcomes as "to analyze [outcome] of [population, disease, risk factor, setting] that are treated with [intervention or comparison]", adding subsequently possible secondary study aims with different outcomes, settings or interventions.

Various factors can drive the choice regarding narrower or broader patient selection and interventions comparison and can greatly influence review completion and the application of tis results.

Research questions can be focused on a specific aspect of the problem; they can exclude populations, or patient groups based on certain characteristics, or they can have a broader spectrum. To decide on the broadness of reviews, researchers must consider the available resources to complete the project, as including more data means an increased workload and ease of use of the results. Broader reviews may have extreme heterogeneity among participants and results; they might be difficult to apply to real-life clinical scenarios but might be more useful in population-based observations or interventions, although heterogeneity and applicability will depend on patient recruitment in the primary studies.

While more focused reviews might be more directly applicable for healthcare professionals, they can't directly provide evidence on differences in outcomes in other settings or populations unless a specific review is carried out. In some cases, a narrower focus means scarce evidence available to summarize and can bring inconclusive results. A much broader review study can be split into more specific research questions, which could also be modified in the project's second phase. These sub-questions can distinguish among patients or interventions (with different PICO questions) to create more homogenous groups that can be more easily compared and can be more beneficial for readers, while providing all available evidence on the topic in one publication. (Caldwell and Welton 2016)

When the studies have been already summarized in a specific review, and the number of primary studies is considerable, creating a new systematic review can be difficult whilst providing a small advancement on what's already been analyzed. In those cases, an umbrella review (overview of reviews) can be useful to summarize the findings available for different interventions for the same condition or for diverse settings; and when comparisons have at least one arm in common network meta analyses can be considered.

In summary, a research question is a milestone to start a review project; it has to be well discussed among stakeholders and researchers at the beginning to set the right path for the rest of the review steps. A simple mnemonic to use to remember fundamental parts of a good research question/aim is the PICO (Population, Intervention, Comparison(s) and Outcome) method. Broader or narrower review questions can result in fundamental changes in the review execution and result; they should be discussed and decided based on resources, target population and available evidence, although in some cases, additional questions can be added to a bigger ongoing project.

CRITERIA FOR A ROBUST RESEARCH QUESTION

A well-constructed and complete research question might still be of limited impact if the project is much more difficult than the available resources or the idea is not adequately circumstantiated in the needs of end users. Hulley and Cummings (2007) proposed as a guideline to identify strong research questions, that have been identified as the FINER criteria, starting from the acronym of the words: Feasible (materially possible to achieve), Interesting (appeals researchers and stakeholders), Novel (there is a need for new evidence on the topic), Ethical (questions should be framed and prioritized according to ethics and considering the consequences of the research publication), and Relevant (the paper may contribute to change the clinical practice). (Hulley, Cummings et al 2007)

Feasibility

A review is feasible when the project workload is adequate for the economic, human and time resources available for the team. Choices regarding the breadth of the review research question should consider this aspect to allow for a good balance between the ideal amount of detail of the review results and the limits of actual research projects. A rapid search of the current knowledge on the matter, the type of evidence available, and its quantity can be very useful in informing and guiding this decision process. The goal is to find a gap of knowledge in the available literature that can be answered through an analysis of current data (without incurring the cases of "empty" reviews) and without embarking on a project entailing the review of millions of papers.

Interest

A research question should start from an interest in finding more about a real problem of both researchers (that pushes them to continue working on the topic) and stakeholders (that sustains investigators to move forward with the work to discover its results). Thus, the review should be constructed around the need for data to apply to real-world scenarios to improve current practices. If a review is well constructed and rigorously executed but can't be applied to real scenarios, the work of the involved researchers can't influence health outcomes. It is worth noting that in the real world, even with the universal availability of interesting and relevant results, the widespread application of recommendations obtained from research results is far from a straightforward process that relies on reporting, dissemination, and institutional uptake.

Novelty

When planning to invest resources in a review, a preliminary analysis of current evidence should explore current secondary studies, including their focus and datedness, ongoing studies with registered protocols, guidelines and textbooks in use and expert opinions on important areas of lack of knowledge. A thorough investigation allows researchers to get to know the wider context and use resources effectively without duplication of established knowledge or embarking in a project currently being carried out by colleagues elsewhere. Apart from medical evidence databases, researchers should search the PROSPERO register of systematic reviews and ask stakeholders and clinicians.

Ethical Aspects

Ethics should be taken into consideration in every step of the review creation, from the idea to the elaboration of PICO questions, especially when researchers dialogue with stakeholders (including patients or recipients of possible policy changes). Even though reviews don't directly entail patient interaction or interventions, to carry out these projects, resources must be invested; thus, it is paramount to invest them effectively in projects that can change health outcomes. Priority and framing of questions should always follow ethical directions to respond adequately to patient needs. The analysis of a problem should be reported in the most neutral ways. Nonetheless, that is often difficult with politically loaded topics, and great attention should be placed to avoid unwanted sequelae that can negatively affect involved patients or populations.

Relevance

To produce evidence that can be useful for involved stakeholders and create a strong basis for clinical or public health practices, the review must be relevant to the application in the field. Stakeholder involvement is fundamental to understanding if a review could be useful outside the publication itself. In the preliminary analysis of the literature and study of current practice, the researchers also should find an indication of effect size that can have a clinical relevance that prescinds the possible highly statistically significant differences. The clinical significance regards changes in clinically relevant outcomes, i.e. differences in blood pressure of 1 mmHg, disease incidence changes of 5 people every 10 million and many more possible results that can't warrant a change in clinical practice. Faithful and clear reporting of results is paramount to inform clinicians, patients, and other researchers about the real clinical relevance of the results. The GRADE framework, which indicates how to report review results to facilitate translation into answers to health questions and indications for clinical practice, can also be useful in the designing phase of a review. (Neumann et al., 2015)

Since the end users are fundamental to consider when deciding which review question to answer and how researchers should consider qualitative research review and synthesis, among other means of providing high-quality data. They are the means to explore the experience and subjective opinions of interventions and practice, and they can greatly inform the best practices to offer, especially if quantitative evidence is scarce or difficult to produce and analyze.

Adverse effects of the interventions are as important as the intended effects; they should be considered when designing data search and extraction and later in data synthesis, especially if they are missing from a part of the included evidence, as it can be due to reporting bias or errors that should be underlined in a systematic review. Qualitative studies or longer-term observational studies can present data on adverse events that are not available in trials. Thus, they should be considered at least in the preliminary analysis of current evidence. (Kneale et al., 2015)

Possible biases in reporting and inadequate availability of evidence should be explored, especially when formulating a study protocol, to account for the systematic error in the results. Skewed results should be recognized, and their causes analyzed to avoid one-sided syntheses of evidence and perpetuating the bias present in the literature, especially when strong economic pressure is present. Especially in those cases, conflicts of interest should be taken into consideration in data extraction synthesis and presentation.

When selecting robust research questions from a clinical setting, it can be difficult to find the perfect balance between what is important and what can be achieved through a feasible review. A good starting point would be to analyze clinical questions on the chosen topic. A broad question would be how to reduce morbidity or mortality for patients with pancreatitis; on the contrary, a hyper-specific question would be how to manage fluid intake for a specific 80-year-old patient with pancreatitis, considering their comorbidities, current clinical condition and needs, and the hospital's available resources. To achieve a realistic research project that can give applicable answers to clinical gaps of knowledge, it is necessary to identify a balance between broadness and detail in the research questions.

It can be helpful to write explicitly the components of the question, beginning from identifying the focus area (or "type of clinical question" that is usually divided into: Diagnosis, Therapy, Prognosis, Harm/Etiology) and defining population, intervention, comparison(s) and outcome. For example, a clinical researcher who is treating patients with pancreatitis can have questions on the therapy and can ask: adult patients admitted with acute pancreatitis (*patients*) should be treated with a large amount of fluids (*intervention*) or a moderate amount of fluids (*comparison*) to achieve lower mortality rates within 90 days (*outcome*). If multiple questions arise, one can choose where to begin from the most pressing doubts present in clinical practice and from the more prominent gaps in current literature.

The process entails multiple options and trials that should go through team discussions and stakeholders' approval following the FINER criteria guidelines. From an initial question, the researcher must go through a thorough selection and refinement process to obtain a feasible project, analyzing the current literature to center the gaps of knowledge and the number, topic and quality of currently available studies that could be reviewed, and achieving a synthesis that can answer the clinicians and patients' needs in real world scenarios, potentially changing clinical practice.

In summary, a good research question shouldn't only be crafted clearly and completely, following the PICO criteria to reach an adequate definition of the study aim but also well circumstantiated in the current research panorama. The FINER (Feasible, Interesting, Novel, Ethical, and Relevant) criteria are a simple and effective guide for researchers to craft research questions that are useful to end users and stakeholders and can potentially change clinical practices.

CONSIDERING STAKEHOLDERS AND END USERS

Stakeholders are people or groups, associations, and institutions that intervene or are affected by health and healthcare-related decisions informed by research (Concannon, 2012). When including the category of stakeholders in a research planning setting, the term is purposely kept as broad as possible to identify all possible people involved in the setting. They might be unclear at the very beginning of the project, but in the design process, they should always be identified and involved in reaching a consensus. This broad definition can consist of patients, caregivers, patient associations, the wider population that can be potentially affected, ethical boards, health care providers, epidemiologists/public health departments, institutions that provide health care and finance health, ministries and policy makers, research institutions, research funders, medical publishers, pharmaceutical or medical device industries.

Multiple research has noted that greater diversity in involved parts regarding knowledge, background and goals can increase skills available for decision-making and improve the overall quality of the research works. This task could be challenging and require time, as increased diversity means different goals and priorities and can bring difficulties in finding a shared opinion on the research question. (Oliver, 2018)

Priorities that were already discussed and crafted into position papers or deliberations from an international to a local level can be a helpful start to align to and begin a focused discussion on the details a specific research might aim to inform on. (Nasser 2018) It can be useful to analyze previous papers, guidelines or opinion letters regarding advice for future research areas to confirm the goals that current researchers advise to focus on. Priority-setting through previous research analysis can help researchers have a useful guideline of relevant, interesting and ethical research ideas to cover knowledge gaps identified by stakeholders and, therefore could be more useful in real-life scenarios than others (Chalmers et al 2014, Bryant et al 2014, Tong et al 2015)

Once a detailed PICO-based research question is crafted, stakeholders should be involved to check if the researchers' interpretation of the priorities was adequate and if the question can satisfactorily answer a pressing clinical question as a broader problem can have numerous nuances. The decision on the review question broadness can be answered in this second phase of research planning, based on feasibility of the project and advice from involved parties (i.e., excluding specific populations, therapies, settings etc. or creating subgroups during data analysis).

A comprehensive assessment and involvement of stakeholders, even if time-consuming, should always be followed as it can have a great impact on planning and outcomes of research. Especially when the context of the review wasn't already established through priority setting analysis or the results are meant for a specific group (a guideline aimed to community health workers or meant to be applied in a specific country) it's necessary to involve various stakeholders to inform review planning. (Magwood et al., 2022)

Creating meaningful results aims to amplify the invested resources to increase health outcomes. The limited resources allocated for research are often decided by external financers that can understand in a limited manner the more pressing problems and the research models more indicated to solve them, the projects are carried by researchers that often time won't apply directly the results, or the guidelines informed by the results. That is why priorities can't be decided uniquely by one of these parties but should be chosen by informed consensus to avoid a splurge of material resources or time.

It's often difficult to identify the project to invest in to have the "most effective" spending as stakeholders often bring differences in priorities; research institutions invest more easily into basic research or pharmacological advances, and it's less common to spend on repeat experiments or on the translation of results across models. Instead patients and end users of the studied scientific advances would prefer to know more on the effects of interventions for the functional, social, and emotional wellbeing, drugs adverse reactions and long-term disease outcomes. Even the most ideal research design can be slowed down or halted by methodological disincentives and difficulties. Designing, running and interpreting research with clear and easily measurable outcomes is easier than including quality of life, psychological or physical subjective assessments and opinions on care and long-term expectations and results. These results are often measured separately in non-experimental models creating different levels of quality of evidence. In reviews, these measures are difficult to analyze and unify in single measurements, and the use of different models and questionnaires makes it difficult to appraise these tools compared to "hard outcomes". (Lorenc et al., 2013, Nasser et al., 2013)

From an initial research aim and relative question, one can start wondering how to identify and engage stakeholders that are relevant to the problem at hand and the possible solutions. A good question would be "Who is the end user that will be affected by the final health outcome and who are the parties involved to make that outcome happen in the real world (patient advocates, physicians, societies that write guidelines, community workers)?". These stakeholders are fundamental to enable the research result to be meaningfully applied and for funds to be invested in efforts that are interesting to them. In the past, research projects were discontinued after discussions with the public, as their focus was outside the main priorities (Boote et al., 2012; Oliver et al. 2009). End-users of systematic review of basic research are scientists themselves, institutions, and funders who should decide on new paths for basic scientific research.

On the other end, it's important to consult who would invest money to fund the research efforts and to apply the results in hospitals or communities if it's known that a considerable amount of money or a specific device is virtually unavailable in the context that is studied the results won't be helpful at the time being and research could be redirected. Researchers should be interested in bringing forward the research efforts and advocating for the results to be applied and spread (not only in the research panorama but to all stakeholders). Ideally stakeholders should be involved from the review conception to its reporting and dissemination also to the greater public. However, in many projects continuous contact and discussion is not always feasible and strategic timings for consultations are instead chosen.

The stakeholder's role is fundamental to avoid as much as possible the known paradoxical effects of research, bringing a harmful effect to recipients, from drug testing to research applied to community behavior. An overview of potential harms can inform clinicians for decisions and future research to avoid these when planning future interventions. Identifying potential harms of an intervention during planning can help measure them in the review, especially if they occur so rarely to be detected only when large patient numbers are included. It could uncover reasons of different reporting across settings. Condensing all literature in a review can also help to analyze their mechanisms, which might be partially known beforehand and partially might be discovered in post hoc investigations. Qualitative research often complements qualitative data in these cases, and it should be considered for inclusion to have a more comprehensive outline of the problem in a systematic review. (Bonell et al, 2015)

In summary, stakeholders should be involved in various moments during review planning to avoid the waste of research funds and to aim for the best health outcomes with the available resources. Moreover, they can help identify shortcomings of research, possible harmful effects and inapplicable results.

ADAPTING TO EMERGING TRENDS

To obtain a research focus that holds relevance for stakeholders, researchers must adapt to the most updated gaps in knowledge and research interests, that is especially true for systematic reviews as they focus on data that has been recorded in the preceding years and current practices must be taken into consideration when utilizing older evidence to inform recommendations. Recognizing these trends is critical for researchers and should be considered while formulating the scope and research questions. For example, when a new disease appears in a different area or poses new challenges, as happened during the COVID-19 pandemic, relying on, and synthesizing known information was fundamental to accelerate research on the topic. Participating in discussions, journal clubs, congresses, and meeting with stakeholders can help the researcher to develop a clear idea of what is interesting and relevant in the current research panorama. Interacting with experts in the field can also give the researcher an idea of how previous novel areas were explored and discovered and what is still missing. Although it's important to avoid repeat research and dispersion of funds, and that should be held into consideration evaluating registered research protocols before designing a novel work. While each health research discipline seems to have its distinct knowledge gaps and research interests, certain emerging trends are relevant across the board.

Asking a Good Question Utilizing a Multidisciplinary Approach

As a general recommendation, an interdisciplinary approach can be an interesting starting point for a novel and relevant research focus. By combining insights and approaches from diverse domains, researchers can unlock solutions and views that would be impossible to achieve with a single-discipline lens. In the past, numerous researchers utilized this approach to deliver many creative ideas, combining two or more topics that seemed unlinked or applying the same model to describe and analyze complex data. For example, phylogenetic trees or linguistic tree structures with interconnected tree branches use the same graphic idea to unravel connections between different entities, while being completely different scientific settings. Another example is adapting biomimicry in architecture with the BUGA Fibre Pavilion in Heilbronn (Germany) (Sonntag et al., 2019), where architects emulated the biological load-bearing structures and beetle wings and made the pavilion five times lighter than a conventional steel structure without using molds or cores, minimizing production waste. Surprisingly, even complex unrelated subfields can hold potential for fruitful combinations.

The examples provided, while fascinating, are not that far from everyday applications in scientific production. In well-researched areas demanding innovation, the incorporation of a multidisciplinary approach can prove to be highly beneficial. This approach might be less fruitful in any niche with only a few published articles, as specific research is still needed, and any new discovery remains relevant within the niche. In contrast, in more saturated areas of research with a great number of new manuscripts being crafted each year, adopting this approach can help narrow the scope and render reviews more feasible, and pave the way for creative innovation, by combining the concepts that seemed very distant among each other.

An interdisciplinary approach brings greater depth and comprehensiveness, which is imperative for a pragmatic and useful systematic review. Research problems rarely exist purely in a niche area of research. By incorporating insights from other disciplines, researchers can gain a richer understanding of the factors at play. Imagine studying the effectiveness of a new exercise program for weight loss. A single-discipline approach might focus solely on physiological changes. However, an interdisciplinary approach could consider psychological factors like motivation or sociological aspects of social support, leading to a more comprehensive research question and possible solution, reducing other possible confounders.

In general, asking a research question adopting an interdisciplinary approach also helps to avoid tunnel vision and address real-world scenarios in the long run. Many real-world challenges are now inherently interdisciplinary (such as climate change, health equity, or urban planning). Solving them requires collaboration across fields. Asking comprehensive and precise questions is the very first step in this journey. Disciplinary silos can lead scientists to solve only a section of the problem, where researchers focus narrowly on their own field and confront themselves only with colleagues from the same discipline. This can hinder comprehensive problem-solving and implementation of results for end-users. A mindful focus on asking interdisciplinary questions force us to step outside the known areas and comfort zones and consider alternative viewpoints, which allows to tackle challenges more effectively.

Integrating Artificial Intelligence in Research

Another topic that is often among novel trends is artificial intelligence (AI) integration, that can be employed from patient care to data collection and analysis. The goal is to use new technology to streamline longer processes that can hinder research development. In the review setting, AI can be integrated into most steps, including the development of focus and research question. In 2013, in honor of 20 years of meta-analysis and evidence synthesis, Doug Altman stated unequivocally that two major areas to be addressed in the future should be reducing the time and effort required to conduct a Cochrane Review without compromising rigorousness and presenting results diversely suitable for different readerships (Altman, 2013). In light of this, future applications of AI, specifically large language models (Gillman & Wright, 2010), appear to be a useful tool to aid researchers to be up to date. Since a few years ago, it has effectively accelerated healthcare research. These tools are extensively and creatively integrated throughout the research process; however, there are still significant concerns and ethical debates on this matter.

The technology has come a long way since its inception, from rule-based systems to complex models powered by deep learning. The first Generative Pre-trained Transformer model (GPT-1), introduced in 2018, was a groundbreaking model capable of generating human-quality text. Subsequent iterations like GPT-2 and GPT-3 increased in size and sophistication, with GPT-3 reaching a staggering 175 billion parameters and achieving near-human-generated content. In 2022, GPT-3.5 powered ChatGPT was able to engage in human-like conversations, while 2023 brought GPT-4, a multimodal model that can not only process text but also understand images. This progression reflects the significant impact of deep learning on language models, showcasing the potential to revolutionize various applications (Wu et al., 2023).

These models aid in knowledge exploration, brainstorming, idea generation, and research question refinement when formulating research questions for a systematic review and manual labor can be significantly reduced (Eloundou, et al., 2023; Kausik, 2023). Imagine a team that is reviewing the effects of mindfulness meditation on anxiety. They may usually start the preliminary search on databases such as PubMed or

Scopus and refine their idea manually trough a longer brainstorming. Alternatively, GPT-based tools could be instructed to search scholarly databases for relevant studies on mindfulness and anxiety. The model could identify knowledge gaps and suggest new research questions by analysing this vast amount of literature. For instance, GPT might reveal a lack of studies on the long-term effects of mindfulness meditation for teenagers with anxiety, prompting the researcher to formulate a question targeting this specific population. Furthermore, GPT can assist with brainstorming and idea generation. The researcher could input existing knowledge on mindfulness and anxiety and ask GPT to suggest related concepts or unexplored aspects, potentially sparking innovative research avenues. Finally, GPT can be a valuable tool for refining research questions using the PICO framework. By prompting GPT with Population (teenagers with anxiety), Intervention (mindfulness meditation), and Outcome (anxiety reduction), the researcher can receive suggestions for Comparison groups (traditional or conventional therapy) and specific Timeframes (e.g., 8-week program) to strengthen their research question. Another significant application of the technology is that it allows non-native English-speaking researchers to present their ideas and translate them effectively or write or correct a text as an effective proofreading tool.

Some even think GPT-based tools can disrupt the publishing culture (Conroy, 2023), although major technical and ethical concerns remain the main barriers to a widespread application and should be a focus for future development. First, the trustworthiness and consistency of many AI models remain questionable, where hallucinations, meaning the generation of false data when missing, seem to be an innate limit (Xu, Jain, & Kankanhalli, 2024). Second, the transparency and interpretability of AI models remain challenging. Many operate as "black boxes," making it difficult to understand their decision-making process and questions about accountability and trust. Third, the human-AI collaboration and its ethics is a persistent debate. It is still unclear whether the amount of AI-based content in research is reasonable.

There is a growing concern that AI will replace human expertise, but finding the right balance—where AI augments human capabilities—is essential. Current approaches from scientific publishers are not consistent, where a few requires transparency if AI-powered tools were used (e.g.,

Nature), while the others ban the use of LLMs (e.g., *Science*) (Conroy, 2023). Global multidisciplinary initiatives that can provide international consensus and guidelines on integrating AI are urgently needed to handle these issues in the near future. There are now plenty of AI-powered tools that streamline the research process. These tools aim to enhance efficiency, reduce manual labor when possible, and improve the quality of systematic literature reviews. There are Covidence, Rayyan, Elicit, SciSpace, DistillerSR, EPPI-Reviewer 6, Abstrackr, and so on. The choice of how and when to use AI tools in research projects depends on personal perspective. Regardless of choice, it's fundamental to remember that while these tools can be valuable, researchers should always exercise critical judgment and validate the outputs. They should complement human expertise rather than replace it. Adhering to established reporting guidelines (such as PRISMA) and other common practices ensures transparency and rigor in systematic reviews.

In conclusion, emerging trends should be analyzed by researchers to understand how to implement their novel idea into topics of interest for stakeholders and end users and current scientist. Every discipline has different trend that can vary a lot in time, some common trends are multidisciplinary approaches that can be useful to confer new ideas. Another new trend is AI that can be employed in research projects from the planning to the application of the project.

Balancing Specificity and Flexibility

A good set of search strategy and eligibility criteria, that are at the core of every systematic review, is founded on a good scope and research questions following the PICO framework. A research focus should be well specified and formulated keeping in mind the application of results in the real-world setting, as less specific researches might remain unapplicable. When in doubt on how to choose specific parameters to focus on, useful tools to set the research question limits can be: (i) consulting with subject matter experts in the field, (ii) pilot testing your question by searching for relevant studies in databases, and (iii) consult a librarian to refine the research question, for example using Boolean operators (AND, OR, NOT) in search engines and other advanced search techniques to further refine the search strategy and keep the project feasible.

In formulating research questions, it is crucial to strike a balance between specificity and a degree of flexibility. The process should begin with a clear and shared definition of the PICO components; for example, the population of interest section should include age, gender, diagnosis, or a specific condition. However, considering broader populations is recommended if the initial focus is too narrow and might limit relevant studies. The same principle applies to the *intervention* being studied. While it is important to clearly define the specific techniques and protocols of eligible studies, including broader categories of interventions, such as a class of drugs instead of a specific one, can provide flexibility if the initial focus is too narrow and can avoid ignoring important evidence. The *comparison* group should also be clearly defined, including alternative interventions, control groups, or usual care, and it should be registered in detail during the analysis to spot confounders. Yet, flexibility is key here, too, as a no-intervention or placebo group might be relevant, or even different types of interventions for comparison, the possibility of subgroup and network meta-analysis should be explored. Lastly, the primary and secondary *outcomes* of interest should be clearly indicated, focusing on relevant and measurable indicators. While "hard outcomes" such as mortality can be more reliable, a broader range of potential outcomes, including individual opinions on treatment and quality of life should also be considered to capture both intended and unintended effects. This approach ensures a comprehensive and adaptable research design.

For example, a team starts with a broad search on the topic on breastfeeding during pregnancy (i.e., mothers who bear a baby while breastfeeding another one) (Minh et al., 2021), when exploring the current evidence related to the mother experience (*population*), they found numerous studies discussing aspects such as nutritional impact, mental health, or hormonal changes. However, after a thorough review of the literature, they decided to include the advantages and disadvantages of both the mothers and their fetuses as significant data was available on both, and it was deemed important to integrate it and the current trends and research priorities that put a focus on maternal and fetal nutrition.

Moreover, to further maintain flexibility and avoid the exclusion of relevant research, the researchers can decide to avoid restricting the target population to a specific group (like first-time mothers or mothers of a certain age) at this stage and collect all available evidence. The question concerns the fetomaternal outcomes of breastfeeding during pregnancy; it sounds relevant to the chosen topic while remaining feasible. This question is specific enough to guide research, but also flexible enough to allow the team to explore different populations of breastfeeding mothers. If during the research, this team finds that a particular population (like first-time mothers or mothers of a certain age) is the only one with a good amount of available data or is more relevant to their stakeholders' suggestions, they can then narrow down the population for a more specific research question. This is an example of applying *flexibility* of research ideas based on trends, current research priorities and stakeholders' interest.

Continuing with formulating research questions, the timeframe for the intervention and settings should be clearly defined. Considering a narrower or broader timeframe might offer valuable insights if studies with different durations are relevant depending on the context. The settings where the intervention is delivered, such as a hospital, outpatient clinic, or community, can also affect the search results. While ignoring this data in research question formulation can bring less applicable data in the real world. A broader set of settings should be considered if the initial focus is too narrow and might exclude relevant studies from different environments, and vice versa.

In some peculiar cases, such as new emerging diseases or new technologies, the question and inclusion criteria of the original project might change because new information is available, for example, to include other specific situations, such as new therapies or new diagnostic tools. Input from individuals involved in the field (such as clinicians, patients, and policymakers) can help define a new focus based on evolving circumstances. In other cases, the research methodology can change when needed; for example, when there is a predominance of qualitative data, researchers can choose to synthesize that information or employ new datasets that wasn't fully known at the beginning of the project.

When the research focus and methodology are not well aligned, the evidence the project's produces can be far from the targeted research priorities and different from what is needed by recipients. For example, if a weight loss program analysis doesn't record which studies include individuals utilizing specific drugs or surgical interventions, it can have a significant bias and hold little significance for patients and healthcare workers. While a study on the effect of an extremely new drug on long-term disease prevention can yield inconclusive results if there is scarce data available and conclude ineffectiveness when there is a lack of evidence, and might halt further investments.

In conclusion, striking the right balance between specificity and flexibility is crucial. A question should aim to be specific but researchers should take into consideration the available evidence, including possible scientific advances and if needed alter the proposed project, question or methodology to keep the review interesting and relevant, while maintaining a dialogue with stakeholders. This approach ensures a comprehensive and adaptable research design that will aid in balancing the specificity and flexibility of the review questions.

CONCLUSION

The paramount importance of a clear and well-defined scope, coupled with a precisely formulated research question, cannot be overstated in guiding the entirety of the research process. Establishing a clear scope is a powerful deterrent to ambiguity, ensuring that research objectives remain focused and attainable. Moreover, identifying the scope necessitates a thoughtful consideration of existing knowledge, gaps, and potential contributions to the field, laying the groundwork for impactful research. Adherence to stringent criteria further refines the research question, guaranteeing a focused and meaningful contribution to the chosen field.

Moreover, the incorporation of stakeholder needs and perspectives elevates the research's relevance and impact. Embracing adaptability to accommodate emerging trends and evolving priorities, coupled with a dynamic approach allowing for refinement, maintains the integrity of research objectives. Ultimately, the possibility to adapt to emerging trends and novelty, while maintaining a feasible and coherent project enhances the effectiveness of the research. In summary, a meticulously crafted scope and research question emerge as foundational pillars, guiding every subsequent stage of the research process and proving fundamental to the overall success of the research endeavor.

REFERENCES

Altman, D. (2013). Twenty years of meta-analysis and evidence synthesis methods: A personal reflection. *Cochrane Database of Systematic Reviews*, 2–11.

Barbour, R. S. (2001). Checklists for improving rigour in qualitative research: A case of the tail wagging the dog? *BMJ (Clinical Research Ed.)*, *322*(7294), 1115–1117. DOI: 10.1136/bmj.322.7294.1115 PMID: 11337448

Bhattacherjee, A. (2012). *Social Science Research: Principles, Methods, and Practices*. University of South Florida.

Bonell, C., Jamal, F., Melendez-Torres, G. J., & Cummins, S. (2015). 'Dark logic': Theorising the harmful consequences of public health interventions. *Journal of Epidemiology and Community Health*, *69*(1), 95–98. DOI: 10.1136/jech-2014-204671 PMID: 25403381

Boote, J. D., Dalgleish, M., Freeman, J., Jones, Z., Miles, M., & Rodgers, H. (2012). But is it a question worth asking? A reflective case study describing how public involvement can lead to researchers' ideas being abandoned. *Health Expectations*. PMID: 22646745Booth, A., Sutton, A., & Papaioannou, D. (2016). *Systematic approaches to a successful literature review*. SAGE Publications.

Chalmers, I., Bracken, M. B., Djulbegovic, B., Garattini, S., Grant, J., Gülmezoglu, A. M., Howells, D. W., Ioannidis, J. P. A., & Oliver, S. (2014). How to increase value and reduce waste when research priorities are set. *Lancet*, *383*(9912), 156–165. DOI: 10.1016/S0140-6736(13)62229-1 PMID: 24411644

Conroy, G. (2023). How ChatGPT and other AI tools could disrupt scientific publishing. *Nature*, *622*(7982), 234–236. DOI: 10.1038/d41586-023-03144-w PMID: 37817033

Creswell, J. W., & Creswell, J. D. (2017). *Research Design: Qualitative, Quantitative, and Mixed Methods Approaches*. SAGE Publications.

Eloundou, T., Manning, S., Mishkin, P., & Rock, D. (2023). Gpts are gpts: An early look at the labor market impact potential of large language models. arXiv preprint arXiv:2303.10130.

Gillman, L. N., & Wright, S. D. (2010). Mega mistakes in meta-analyses: Devil in the detail. *Ecology*, *91*(9), 2550–2552. DOI: 10.1890/09-0339.1 PMID: 20957947

Gough, D., Oliver, S., & Thomas, J. (2017). *An Introduction to Systematic Reviews*. SAGE Publications.

Grant, M. J., & Booth, A. (2009). A typology of reviews: An analysis of 14 review types and associated methodologies. *Health Information and Libraries Journal*, *26*(2), 91–108. DOI: 10.1111/j.1471-1842.2009.00848.x PMID: 19490148

Greenhalgh, T. (2014). *How to Read a Paper: The Basics of Evidence-Based Medicine*. John Wiley & Sons.

Hart, C. (1998). *Doing a Literature Review: Releasing the Social Science Research Imagination*. SAGE Publications.

Hulley, S. B., Cummings, S. R., & Browner, W. S. (Eds.). (2007). *Designing Clinical Research: An Epidemiological Approach* (4th ed., pp. 14–22). Lippincott Williams & Wilkins.

Kausik, B. (2023). Long tails & the impact of GPT on labor. Available at *SSRN* 4525008. DOI: 10.2139/ssrn.4525008

Kneale, D., Thomas, J., & Harris, K. (2015). Developing and optimising the use of logic models in systematic reviews: Exploring practice and good practice in the use of programme theory in reviews. *PLoS One*, *10*(11), e0142187. DOI: 10.1371/journal.pone.0142187 PMID: 26575182

Lorenc, T., Petticrew, M., Welch, V., & Tugwell, P. (2013). What types of interventions generate inequalities? Evidence from systematic reviews. *Journal of Epidemiology and Community Health*, *67*(2), 190–193. DOI: 10.1136/jech-2012-201257 PMID: 22875078

Magwood, O., Riddle, A., Petkovic, J., Lytvyn, L., Khabsa, J., Atwere, P., Akl, E. A., Campbell, P., Welch, V., Smith, M., Mustafa, R. A., Limburg, H., Dans, L. F., Skoetz, N., Grant, S., Concannon, T. W., & Tugwell, P. (2022). PROTOCOL: Barriers and facilitators to stakeholder engagement in health guideline development: A qualitative evidence synthesis. *Campbell Systematic Reviews*, *18*(2), e1237. DOI: 10.1002/cl2.1237 PMID: 36911345

Mayer, S. J. (2018). *The Dynamic Student Development Meta-Theory: A New Model for Student Success*. Routledge.

Miles, M. B., Huberman, A. M., & Saldaña, J. (2014). *Qualitative Data Analysis: A Methods Sourcebook*. SAGE Publications.

Minh, L. H. N., Tawfik, G. M., Ghozy, S., Hashan, M. R., Nam, N. H., Linh, L. K., & Nhat Le, T. (2021). Feto-maternal outcomes of breastfeeding during pregnancy: A systematic review and meta-analysis. *Journal of Tropical Pediatrics*, *67*(6), fmab097. DOI: 10.1093/tropej/fmab097 PMID: 34962568

Nasser, M., Ueffing, E., Welch, V., & Tugwell, P. (2013). An equity lens can ensure an equity-oriented approach to agenda setting and priority setting of Cochrane Reviews. *Journal of Clinical Epidemiology*, *66*(5), 511–521. DOI: 10.1016/j.jclinepi.2012.11.013 PMID: 23477991

Neuman, W. L. (2014). *Social Research Methods: Qualitative and Quantitative Approaches*. Pearson Education.

Neumann, I., Brignardello-Petersen, R., Wiercioch, W., Carrasco-Labra, A., Cuello, C., Akl, E., Mustafa, R. A., Al-Hazzani, W., Etxeandia-Ikobaltzeta, I., Rojas, M. X., Falavigna, M., Santesso, N., Brozek, J., Iorio, A., Alonso-Coello, P., & Schünemann, H. J. (2015). The GRADE evidence-to-decision framework: A report of its testing and application in 15 international guideline panels. *Implementation Science : IS*, *11*(1), 93. DOI: 10.1186/s13012-016-0462-y PMID: 27417219

Oliver, S., Armes, D. G., & Gyte, G. (2009). Public involvement influences a national research agenda. *Patient*, *2*(3), 179–190. DOI: 10.2165/11314860-000000000-00000 PMID: 22273169

Patton, M. Q. (2015). *Qualitative Research & Evaluation Methods: Integrating Theory and Practice*. SAGE Publications.

Petticrew, M., & Roberts, H. (2008). *Systematic Reviews in the Social Sciences: A Practical Guide*. John Wiley & Sons.

Richardson, W. S., Wilson, M. C., Nishikawa, J., & Hayward, R. S. (1995). The well-built clinical question: A key to evidence-based decisions. *ACP Journal Club*, *123*(3), A12–A13. DOI: 10.7326/ACPJC-1995-123-3-A12 PMID: 7582737

Smith, J. A. (2015). *Qualitative Psychology: A Practical Guide to Research Methods*. SAGE Publications.

Sonntag, D., Aldinger, L., Bechert, S., Alvarez, M., Groenewolt, A., Krieg, O. D., & Menges, A. (2019). Lightweight segmented timber shell for the Bundesgartenschau in Heilbronn. Paper presented at the Proceedings of IASS Annual Symposia.

Wu, T., He, S., Liu, J., Sun, S., Liu, K., Han, Q.-L., & Tang, Y. (2023). A brief overview of ChatGPT: The history, status quo and potential future development. IEEE/CAA Journal of Automatica Sinica, 10(5), 1122-1136.

Xu, Z., Jain, S., & Kankanhalli, M. (2024). Hallucination is inevitable: An innate limitation of large language models. arXiv preprint arXiv:2401.11817.

Yin, R. K. (2017). *Case Study Research and Applications: Design and Methods*. SAGE Publications.

ADDITIONAL READING

Keown, K., Van Eerd, D., & Irvin, E. (2008). Stakeholder engagement opportunities in systematic reviews: Knowledge transfer for policy and practice. *The Journal of Continuing Education in the Health Professions*, *28*(2), 67–72. DOI: 10.1002/chp.159 PMID: 18521874

Thomas, J., Kneale, D., McKenzie, J. E., Brennan, S. E., & Bhaumik, S. (2023) Chapter 2: Determining the scope of the review and the questions it will address. In: Higgins JPT, Thomas J, Chandler J, Cumpston M, Li T, Page MJ, Welch VA (editors). Cochrane Handbook for Systematic Reviews of Interventions version 6.4 (updated August 2023). Cochrane. Available from www.training.cochrane.org/handbook

KEY TERMS AND DEFINITIONS

FINER Criteria: Standing for "Feasible, Interesting, Novel, Ethical, and Relevant", they are the criteria for a well circumstantiated research question that can bring to a research work that can benefit community and end users of interventions.

PICO Criteria: Standing for "Patient, Intervention, Comparison, Outcome", they are guidelines to include all the necessary components to a robust research question

Research Landscape: It's a term commonly used to refer to the current available evidence that is published (or planned to be executed and published in case of protocols) in the scientific literature

Research Methodology: It refers to the detailed definition of the type of research review employed to achieve a rigorous and relevant result

Research Question: Is a detailed question that researchers use to expose clearly the data they want to collect and analyse to provide an answer to users and other researchers trough their research

Scope: It can be a broader or ample area on which the researchers chose to focus on, when analyzing what is already known and what is missing

Stakeholder: A broad term to identify people or institution that have interest in the topic and results of research, for example they invest money, can benefit form the intervention or are in charge of selecting the best intervention for patients

Chapter 6
Developing the Inclusion and Exclusion Criteria

Khanh Dang Nguyen
https://orcid.org/0009-0001-3184-3778
University of Medicine and Pharmacy at Ho Chi Minh City, Vietnam

Hoang Anh Nguyen
https://orcid.org/0009-0000-5271-7458
Thai Binh University of Medicine and Pharmacy, Thai Binh, Vietnam

Engy Mohamed Makram
https://orcid.org/0009-0006-1129-1183
Misr University for Science and Technology, Giza, Egypt

Toka Adel Hassan
https://orcid.org/0000-0001-6166-6955
October 6 University, Giza, Egypt

ABSTRACT

The initial phase for conducting SR/MA involves systematically searching multiple databases and locating all potentially relevant studies that could contribute to evaluating the review question. The reviewers establish

DOI: 10.4018/979-8-3693-3007-4.ch006

parameters to skillfully control the search process and to specify which studies fit the systematic review's scope and which do not. These parameters are broadly known as "the eligibility criteria." Eligibility criteria can be further divided into inclusion criteria (to rule in studies) and exclusion criteria (to rule out studies). This fundamental step secures a high-quality research study, serving as the guiding principle for investigators to rely on and address their research questions effectively. This chapter reveals the role of inclusion and exclusion criteria in an SR/MA study. Besides discussing techniques for developing a good set of criteria, we will explore the common pitfalls to avoid, ensuring that the choices made during the design phase resonate with the broader context of research credibility.

INTRODUCTION

The initial phase for conducting SR/MA involves systematically searching multiple databases and locating all potentially relevant studies that could contribute to evaluating the review question. The reviewers establish parameters to skillfully control the search process and to specify which studies fit the systematic review's scope and which do not. These parameters are broadly known as "the eligibility criteria." (Meline, 2006).

Eligibility criteria can be further divided into inclusion criteria (to rule in studies) and exclusion criteria (to rule out studies) (Patino & Ferreira, 2018). This fundamental step secures a high-quality research study, serving as the guiding principle for investigators to rely on and address their research questions effectively. This chapter reveals the role of inclusion and exclusion criteria in an SR/MA study. Besides discussing techniques for developing a good set of criteria, we will explore the common pitfalls to avoid, ensuring that the choices made during the design phase resonate with the broader context of research credibility.

UNDERSTANDING THE RESEARCH CONTEXT

One of the earliest steps in conducting research is formulating a precise research question. Undoubtedly, it is the cornerstone of executing a thorough systematic review. On that account, the PICO (population, intervention, control, and outcomes) format is used to make sure the research question is relevant, which helps in developing an appropriate research question (Aslam & Emmanuel, 2010), (Hulley et al., 2013).

PICO is understood as follows:

- It is crucial to accurately identify the population or the problem that needs to be addressed and its demography.
- Additionally, the intervention or exposure to be considered must be clear. For example, if it is a treatment plan, a diagnostic tool, new operation technique.
- In systematic reviews, there is likely a comparator group, especially if the intervention is about treatment. Therefore, that should also be accurate.
- Lastly, the outcome of the intervention is of utmost importance; it should be specific to the research question.

The next step would be to determine the target population and the study design, this will be pivotal in designing the inclusion/exclusion criteria. Certain study designs could be inappropriate to include in a systematic review. Obviously, other reviews in the literature shouldn't be included. Cohort, case-control, cross-sectional, and clinical trial study designs can be included according to the research question if the study design is not vague or non-specific to the research question. Also, the target population in the included studies is essential for understanding how the inclusion/exclusion criteria will be implemented. For instance, if the sample size includes men and women when the research question specifies a certain sex. Naturally, sometimes studies are inconspicuous about their methodology, so that should be taken into account when including them. It may be needed to disclose that in the limitations (Hulley et al., 2013), (Mckenzie et al., 2019).

If your search question is less clinical there are other, similar approaches, you can use to identify your search concepts, these include SPICE and SPIDER.

SPICE: Noticeably, SPICE is more similar to PICO but splits the population into settings and perspectives, which makes it more specific, and replaces the outcome by evaluation to make it more general. SPICE stands for **S**etting, **P**erspective, **I**ntervention, **C**omparison, and **E**valuation (Booth, 2006).

SPIDER: This stands for **S**ample size, **P**henomenon of **I**nterest, Study **D**esign, **E**valuation, and **R**esearch type. SPIDER integrates the research type and the study design which qualifies the study to be more accurate in developing the inclusion/exclusion criteria and also performing the suitable analysis (Cooke et al., 2012).

DEVELOPING INCLUSION CRITERIA

The establishment of clear and well-defined inclusion criteria serves as a foundational pillar, dictating the success and reliability of a systematic review or meta-analysis.

Initiating the process with a broad perspective. A comprehensive inclusion criterion allows researchers to capture a diverse and expansive sample, ensuring that potential studies relevant to the research question are not inadvertently excluded at the outset. However, the broad starting point is just the beginning. As the journey progresses, the need for specificity becomes evident. Researchers must gradually refine the inclusion criteria based on the intricacies of the research question, the available resources, and the practical aspects of data collection. This step-by-step refinement is essential to strike a delicate balance between relevance and feasibility.

On one hand, the criteria must be specific enough to address the intricacies of the research question, ensuring that the selected studies contribute directly to the synthesis objectives. On the other hand, they must remain achievable regarding recruitment and data collection.

This equilibrium between relevance and feasibility is crucial for the success of the systematic review. Overly stringent criteria may result in a limited pool of eligible studies, potentially undermining the comprehensiveness of the synthesis. Conversely, criteria that are too lenient may lead to the inclusion of studies that do not adequately address the research question (Bettany-Saltikov, 2010).

To provide a practical understanding of the inclusion criteria establishment process, it's essential to consider different domains (Table 1):

1. **Population:** In defining the participant criteria, we delve into the demographic characteristics (age, gender, ethnicity, education level, etc.), behavioral characteristics (specific experiences, behaviors, or activities relevant to the study), health and medical conditions.
2. **Disease or condition:** Clearly outline the specific disease or condition under investigation.
3. **Time frame:** Establish the period during which the study was conducted or published, to ensure relevance to contemporary practice and interventions.
4. **Geographic location:** Specify the geographic areas relevant to the research question.
5. **Setting:** Define the settings where the study was conducted, such as hospitals, communities, or specific geographical regions.
6. **Types of Intervention:** The heart of the matter lies in intervention. Here, clarity is paramount. Let's consider following the previous example and conducting a study on diabetic homeless men. A decision needs to be made on whether to include all treatments (Interventions) for diabetes or only specific ones, such as including only those on Insulin treatment.
7. **Types of Comparative Interventions (PICO Format Only):** If we incorporate a comparative intervention aspect, each comparative group's inclusion and exclusion criteria must be meticulously defined. This ensures a nuanced understanding of the varied components within the study.
8. **Type of Studies:** The first crucial criterion revolves around the type of studies to be included. It must align with the research question, e.g., narrow the focus to randomized controlled trials or cross-sectional Studies. This strategic selection ensures that the chosen studies possess the appropriate design to address the effectiveness of the identified intervention adequately.

9. **Language:** Consider the language of publication to determine eligibility for inclusion.
10. **Publication type:** Determine the types of publications to be included, such as peer-reviewed journals, conference abstracts, or gray literature.

Throughout this process, we adhere to the principles of explicitness and stringency. Each criterion is an outcome of a research question and seamlessly aligns with the identified PICO components. This strategic approach facilitates a systematic and focused search and upholds the integrity of the systematic review.

Table 1. An example of developing a set of inclusion criteria for a study on Ischemic stroke mortality in the Middle East

	Inclusion criteria
Population	Adults aged 18 years and older diagnosed with ischemic stroke in the Middle East.
Disease or condition	Ischemic stroke
Time frame	Studies published within the last 10 years (since the past decade)
Geographic location	Studies conducted specifically in the Middle East region, encompassing countries such as Saudi Arabia, United Arab Emirates, Qatar, etc.
Setting	Various settings within the Middle East region, including hospitals, outpatient clinics, or community healthcare facilities, will be included.
Intervention	Using aspirin as a prophylactic measure for individuals with ischemic stroke.
Comparative intervention (PICO format only)	Using clopidogrel as a prophylactic measure for individuals with ischemic stroke.
Outcomes	To compare the all-cause mortality rates among individuals diagnosed with ischemic stroke in the Middle East, providing insights into treatment effectiveness and safety.
Type of studies	Randomized controlled trials (RCTs)
Language	Studies published in English will be considered for inclusion to ensure comprehension by the research team.

Consistency is paramount in the establishment of inclusion criteria. All members of the review team should be aligned in their understanding and application of the criteria. To achieve this, clear documentation and communication within the research team are essential.

Furthermore, transparency in reporting the inclusion criteria is vital for the credibility of the systematic review. A detailed account of the criteria, accompanied by the rationale behind each choice, should be documented in the protocol. This documentation serves as a reference point for the entire research team and provides transparency for external stakeholders and readers of the final synthesis.

DEVELOPING EXCLUSION CRITERIA

Exclusion criteria are defined as characteristics of the study participants that could potentially increase the risk of a biased outcome or hinder the success of the research (Patino & Ferreira, 2018). Researchers must be aware of potential biases related to study designs, populations, or interventions, striving for objectivity throughout the criteria-setting phase (Pigott & Polanin, 2020). Therefore, determining precise features to build up the exclusion criteria is essential for conducting the SR/MA study while the researchers have to filter hundreds of articles to decide which ones are suitable for the study.

A good set of exclusion criteria needs to be comprehensive so that there will be the least confusion for later steps, such as data extraction and analysis. Moreover, it is not only about how to exclude unsuitable data but also how to optimize the data's effectiveness. When developing the exclusion criteria, always ask yourself this question: "What to collect, and what not?". This question will bring up basically ideas about which characteristics of the study participants you want to investigate. Then ask: "Which factors that could affect these characteristics, leading to an unfavorable outcome of the study?". Bearing these two questions in mind, you will define features that should be eliminated to ensure the success of the study.

To simplify the meaning of exclusion criteria, here we suggest a Venn diagram in the image below:

Figure 1. Venn diagram illustrating the meaning of exclusion criteria

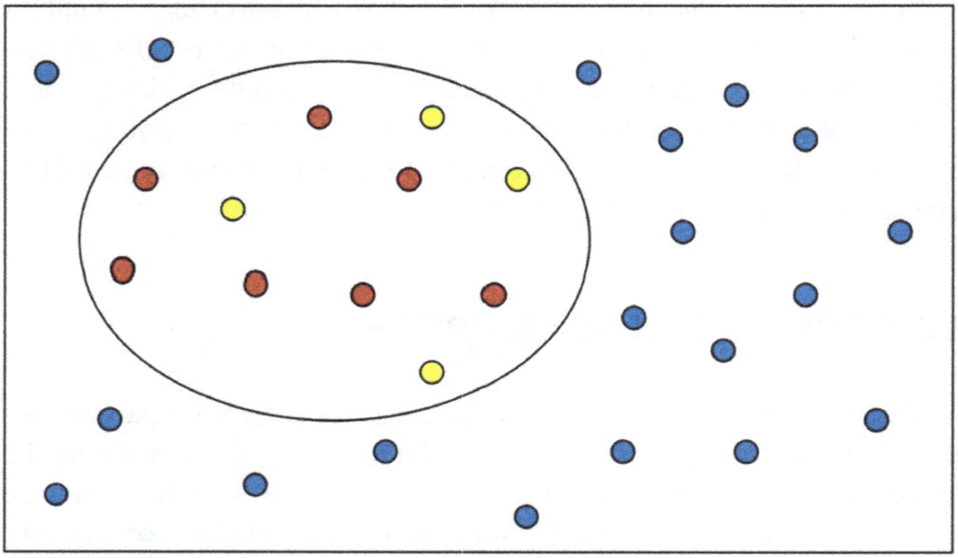

The red dots and yellow dots both represent study participants that are suitable with the Inclusion criteria. However, the yellow ones have features that could lead to a biased outcome (for example: risk factors or protective factors). Blue dots represent study participants that do not match the Inclusion criteria, hence they should be excluded from the study population.

From Figure 1, it is obvious that a good set of exclusion criteria could help the researchers with 2 things: exclude features that are unrelated to the inclusion criteria and limit the unfavorable outcome by eliminating potentially misleading characteristics. These misleading characteristics could be regarded as the ones that increase or decrease the possibility of getting suitable study participants. Imagine that you are conducting a study on the relationship between smoking and oral cancer. You decide to put into the study people who are from 18 – 60 years old, currently smoke, and have this kind of cancer. However, when collecting data from these participants, you realize that about one-third of your participants consume alcohol, which is also a risk factor for oral cancer. The question here is:

will you mention "using alcohol" in your exclusion criteria? Because the possibility of getting participants with oral cancer will get higher when using alcohol together with smoking, this feature should be reconsidered to be excluded from the research population.

Additionally, some important points of exclusion criteria should be noticed by the researchers in all studies. These points aim to not only ensure the success of the study but also to protect the patients during the process of the research. Here are these crucial criteria:

1. Minimized exclusion: justify each exclusion criterion and keep them to a minimum.
2. Ethical exclusion: in some special cases, this would protect the safety of study participants.
3. Factors that could be limitations of the study, such as religion, gender, and age restriction.

In conclusion, developing good exclusion criteria requires researchers to understand profoundly the results they predicted to get. The most efficient way for this step is firstly building a strictly detailed and comprehensive set of inclusion criteria. Then the exclusion criteria will be "the leftovers", which means they will be from features that are not included in the study. Finally, improving exclusion criteria by adding characteristics that potentially mislead the outcome.

TIPS FOR EFFECTIVE CRITERIA DEVELOPMENT

To effectively formulate inclusive and exclusive criteria, in this section, we will discuss some tips and tricks that can assist you in forming the criteria for your research.

- **Clearly Define the Research Question:** The research question serves as the foundation of your systematic review. It should be clear, specific, and focused to guide the development of your inclusion and exclusion criteria. Take time to refine your question to ensure it captures what you want to investigate.

- **Be Specific but Flexible:** While it's essential to have clear criteria, it's also important to remain flexible. As you conduct the review, you may encounter unexpected findings or realize the need to adjust your criteria slightly. Allow for some flexibility while maintaining clarity and focus.
- **Consider Study Design:** Determine which study designs are appropriate for addressing your research question. Common study designs include randomized controlled trials (RCTs), cohort studies, case-control studies, qualitative research, etc. Exclude study designs that are not relevant to your question.
- **Specify Population Characteristics:** Define the demographic and clinical characteristics of the population you are interested in studying. This might include age, gender, ethnicity, health status, geographic location, etc. Be clear about who is included and excluded based on these characteristics.
- **Detail Intervention and Comparator:** Clearly describe the intervention or exposure being evaluated in the included studies. Also, specify any comparators or control groups against which the intervention is being assessed. This helps ensure consistency and comparability across studies.
- **Clarify Outcome Measures:** Define the specific outcomes or endpoints you are interested in assessing. These could be clinical outcomes (e.g., mortality, disease progression), patient-reported outcomes (e.g., quality of life, satisfaction), economic outcomes (e.g., cost-effectiveness), etc. Be explicit about the types of outcomes you want to include.
- **Consider Language and Publication Status:** Decide whether to include studies published in languages other than English and whether to include unpublished studies (grey literature). Depending on the resources available and the scope of your review, you may choose to include or exclude studies based on language and publication status.
- **Consult Stakeholders:** Engage stakeholders such as clinicians, researchers, patients, and policymakers in the development of your criteria. Their input can provide valuable insights into what is im-

portant and relevant for the review, ensuring its applicability and usefulness.

- **Pilot Test Criteria:** Before conducting the full review, pilot test your criteria on a small sample of studies. This helps identify any potential issues or challenges with the criteria and allows you to refine them before proceeding with the full review.
- **Document Decision Rules:** Document the decision rules for applying the criteria during the screening process. Decision rules outline how criteria will be interpreted and applied consistently by reviewers. This promotes transparency and reproducibility in the review process.
- **Review and Revise Criteria:** Periodically review and revise your criteria as needed throughout the review process. New evidence, insights, or stakeholder feedback may necessitate adjustments to the criteria to ensure the review remains relevant and rigorous.

TOOLS AND RESOURCES

By utilizing these tools and resources, you can develop robust and transparent inclusion and exclusion criteria for systematic reviews, enhancing the quality and reliability of your research. Below are some tools that you can try to develop your criteria.

- **Cochrane Collaboration:** The Cochrane Collaboration offers a variety of resources and templates for systematic review development, including guidance on formulating inclusion and exclusion criteria (Scholten et al., 2005).
- **PRISMA-P (Preferred Reporting Items for Systematic Review and Meta-Analysis Protocols):** PRISMA-P provides a checklist and explanation for reporting systematic review protocols, which can help ensure clarity and transparency in defining inclusion and exclusion criteria (Shamseer et al., 2015).
- **PICOTS Framework:** Similar to PICO(T), the PICOTS framework (Population, Intervention, Comparator, Outcome, Timing, Setting)

can be useful for structuring inclusion and exclusion criteria, particularly in healthcare research (Eriksen & Frandsen, 2018).

- **GRADE (Grading of Recommendations Assessment, Development and Evaluation):** GRADE offers guidance on rating the quality of evidence and strength of recommendations in systematic reviews, which can inform the development of criteria (*GRADE Handbook*, n.d.).
- **EPHPP (Effective Public Health Practice Project):** EPHPP provides tools and guidelines for assessing the quality of research evidence, which can inform decisions about inclusion and exclusion criteria (Thomas et al., 2004).
- **ICMJE (International Committee of Medical Journal Editors) Recommendations:** ICMJE offers ethical guidelines for publication in medical journals, which can help ensure that inclusion and exclusion criteria are fair and appropriate (Teixeira da Silva, 2020).
- **PubMed Search Strategy Builder:** PubMed offers a Search Strategy Builder tool to help researchers construct effective search strategies, which can inform the development of inclusion and exclusion criteria.
- **Covidence:** Covidence is a web-based tool for systematic review management, including screening and data extraction, which can streamline the process of applying inclusion and exclusion criteria (*Covidence - Better Systematic Review Management*, n.d.).
- **Rayyan:** Rayyan is another web-based tool for systematic review management, offering features for collaboration, screening, and data extraction, which can facilitate the application of inclusion and exclusion criteria (*Rayyan – Intelligent Systematic Review - Rayyan*, n.d.).
- **Critical Appraisal Tools:** Various critical appraisal tools are available for assessing the quality of different types of research studies, which can inform decisions about inclusion and exclusion criteria based on methodological rigor (*Critical Appraisal Tools*, n.d.).

CONCLUSION

In summary, the development of inclusion and exclusion criteria is crucial for ensuring the integrity and effectiveness of systematic reviews and meta-analyses (SR/MA). Inclusion criteria should capture a diverse range of relevant studies while maintaining specificity to balance relevance and feasibility. Considerations across study types, participant characteristics, interventions, and outcomes guide the establishment of inclusion criteria. As for exclusion criteria, it filters out potential biases and confounding factors. Transparency in reporting and documentation of criteria is paramount for the credibility of the systematic review. Clear communication promotes consistency and enhances the reproducibility of the review process. Lastly, the provided tips and tricks offer practical guidance for developing robust inclusion and exclusion criteria in systematic reviews. They promote clarity, flexibility, and stakeholder engagement, ensuring the relevance and rigor of the review process.

REFERENCES

Aslam, S., & Emmanuel, P. (2010). Formulating a researchable question: A critical step for facilitating good clinical research. *Indian Journal of Sexually Transmitted Diseases and AIDS*, *31*(1), 47–50. DOI: 10.4103/0253-7184.69003 PMID: 21808439

Bettany-Saltikov, J. (2010). Learning how to undertake a systematic review: Part 1. *Nursing Standard (Royal College of Nursing (Great Britain): 1987)*, *24*(50), 47–55; quiz 56. DOI: 10.7748/ns2010.08.24.50.47.c7939

Booth, A. (2006). Clear and present questions: Formulating questions for evidence based practice. *Library Hi Tech*, *24*(3), 355–368. DOI: 10.1108/07378830610692127

Cooke, A., Smith, D., & Booth, A. (2012). Beyond PICO: The SPIDER tool for qualitative evidence synthesis. *Qualitative Health Research*, *22*(10), 1435–1443. DOI: 10.1177/1049732312452938 PMID: 22829486

Covidence—Better systematic review management. (n.d.). Retrieved April 5, 2024, from https://www.covidence.org/

Critical Appraisal tools. (n.d.). [Web Page]. Retrieved April 5, 2024, from https://www.cebm.ox.ac.uk/resources/ebm-tools/critical-appraisal-tools

Eriksen, M. B., & Frandsen, T. F. (2018). The impact of patient, intervention, comparison, outcome (PICO) as a search strategy tool on literature search quality: A systematic review. *Journal of the Medical Library Association: JMLA*, *106*(4), 420–431. DOI: 10.5195/jmla.2018.345 PMID: 30271283

GRADE Handbook. (n.d.). Retrieved April 5, 2024, from https://training.cochrane.org/resource/grade-handbook

Hulley, S. B., Cummings, S. R., Browner, W. S., Grady, D. G., & Newman, T. B. (2013). *Designing Clinical Research*. Lippincott Williams & Wilkins.

Mckenzie, J., Brennan, S., Ryan, R., Thomson, H., Johnston, R., & Thomas, J. (2019). Chapter 3: Defining the criteria for including studies and how they will be grouped for the synthesis. In *Cochrane Handbook for Systematic Reviews of Interventions*. DOI: 10.1002/9781119536604.ch3

Meline, T. (2006). Selecting Studies for Systemic Review: Inclusion and Exclusion Criteria. *Contemporary Issues in Communication Science and Disorders*, *33*(Spring), 21–27. DOI: 10.1044/cicsd_33_S_21

Patino, C. M., & Ferreira, J. C. (2018). Inclusion and exclusion criteria in research studies: Definitions and why they matter. *Jornal Brasileiro de Pneumologia : Publicaçao Oficial da Sociedade Brasileira de Pneumologia e Tisilogia*, *44*(2), 84. DOI: 10.1590/s1806-37562018000000088 PMID: 29791550

Pigott, T. D., & Polanin, J. R. (2020). Methodological Guidance Paper: High-Quality Meta-Analysis in a Systematic Review. *Review of Educational Research*, *90*(1), 24–46. DOI: 10.3102/0034654319877153

Rayyan – Intelligent Systematic Review—Rayyan. (n.d.). Retrieved April 5, 2024, from https://www.rayyan.ai/

Scholten, R. J. P. M., Clarke, M., & Hetherington, J. (2005). The Cochrane Collaboration. *European Journal of Clinical Nutrition*, *59*(S1, Suppl 1), S147–S149. DOI: 10.1038/sj.ejcn.1602188 PMID: 16052183

Shamseer, L., Moher, D., Clarke, M., Ghersi, D., Liberati, A., Petticrew, M., Shekelle, P., Stewart, L. A., & the PRISMA-P Group. (2015). Preferred reporting items for systematic review and meta-analysis protocols (PRISMA-P) 2015: Elaboration and explanation. *BMJ, 349*(jan02 1), g7647–g7647. DOI: 10.1136/bmj.g7647

Teixeira da Silva, J. A. (2020). The ICMJE recommendations: Challenges in fortifying publishing integrity. *Irish Journal of Medical Science*, *189*(4), 1179–1181. DOI: 10.1007/s11845-020-02227-1 PMID: 32297306

Thomas, B. H., Ciliska, D., Dobbins, M., & Micucci, S. (2004). A process for systematically reviewing the literature: Providing the research evidence for public health nursing interventions. *Worldviews on Evidence-Based Nursing*, *1*(3), 176–184. DOI: 10.1111/j.1524-475X.2004.04006.x PMID: 17163895

Chapter 7
Screening Articles for Inclusion in SR/MA

Toka Adel Hassan
https://orcid.org/0000-0001-6166-6955
October 6 University, Egypt

Lam D. Chau
https://orcid.org/0000-0002-1215-087X
Department of Biochemistry, Case Western Reserve University, USA

ABSTRACT

Tailored for junior and beginner researchers in the medical research field, the chapter delves into the crucial process of screening articles for inclusion in systematic reviews and meta-analyses. This chapter serves as a comprehensive guide, addressing the fundamental aspects of selecting articles for inclusion in systematic reviews and meta-analyses. From exploring methods for establishing clear inclusion criteria to navigating challenges and outlining best practices, it equips researchers with the tools needed for a meticulous and reliable screening process to achieve impactful systematic reviews and meta-analyses.

DOI: 10.4018/979-8-3693-3007-4.ch007

INTRODUCTION

The process of screening articles for inclusion is a vital step in high-quality studies. A step-by-step screening process ensures a systematic and error-free synthesis of evidence, through sifting the vast pool of identified articles and allowing only those relevant to the research question to progress through subsequent stages of synthesis.

Central to the screening process is the concept of eligibility criteria, with its two main components " Inclusion Criteria" and "Exclusion Criteria". By employing eligibility criteria, researchers establish a systematic framework for article selection, ensuring consistency and objectivity. These criteria, such as demographic, clinical, and geographic characteristics, serve as predefined standards to determine the relevance and suitability of articles for addressing the research question, and directing the screening process toward articles that align with the research objectives (Meline, 2006; Bettany-Saltikov, 2010). Moreover, systematic reviews incorporate the concept of sensitivity analysis, offering a transparent and comprehensive strategy to handle uncertainties in evidence selection. This strategic approach involves re-evaluating the impact of including or excluding borderline studies, providing an understanding of the robustness and reliability of the evidence synthesis findings (Livoreil et al., 2017).

As the study objectives become more defined, the criteria undergo a process of gradual refinement to enhance specificity, aligning them closely with the nuanced research questions and objectives. This delicate balance between relevance and feasibility is paramount; inclusion criteria must be tailored to effectively address the research question while remaining achievable in terms of participant recruitment and data collection (Patino and Ferreira, 2018).

The approach adopted for eligibility is to apply at least two filters in the screening process. It commences with checking titles and abstracts, allowing for a preliminary assessment of relevance, and then advances to a more in-depth evaluation through a full-text examination. The initial screening for surface-level relevance, followed by a more comprehensive examination of the full-text version grants passage to articles that undoubtedly meet the pre-defined inclusion criteria while subjecting those of ambiguous relevance to further and more discerning inspection (CEE, 2013).

Handling unclear eligibility adds another layer of nuance to the screening process. When faced with ambiguity, a righteous approach involves seeking additional information or clarification (Frampton et al., 2017). This not only allows for a more informed decision but also demonstrates a commitment to thoroughness and accuracy. The screening process, therefore, is not a rigid yes-or-no determination but a nuanced journey of exploration. This comprehensive approach aims to shed light on how a meticulous screening strategy, coupled with well-defined eligibility criteria, contributes to the robustness and reliability of evidence synthesis in research.

PILOT TESTING OF ELIGIBILITY CRITERIA

Pilot testing the eligibility criteria and process is a pivotal phase in the methodological journey of evidence synthesis. This crucial step involves subjecting a sample of articles to the pre-defined eligibility criteria to ensure their accuracy and effectiveness. By meticulously examining around 10-12 articles, including those considered eligible, ineligible, and doubtful, the research team gains insights into the consistency and reliability of their screening process. This pilot testing not only serves as a training ground for the review team to interpret and apply eligibility criteria consistently but also provides an opportunity to identify any unanticipated issues that may have been overlooked during the planning phase. If inconsistencies or challenges arise during pilot testing, it prompts necessary revisions to the eligibility criteria or the application process, fostering an iterative approach toward methodological refinement. The final set of eligibility criteria, validated through pilot testing, is then solidified in the evidence synthesis protocol, laying the groundwork for a systematic and robust article screening process (Green and Higgins, 2011).

ADAPTATION OF INCLUSION CRITERIA TO STUDY OBJECTIVES

Adaptation of inclusion criteria to study objectives is an important process where researchers make decisions on the types of studies for inclusion to the current research of interest. For typical inclusion criteria, it should include all components of PICO(S) (for more information, please refer to **Chapter 8: Developing the inclusion/exclusion criteria**). Adaptation of inclusion criteria to study objectives serves as the groundwork for medical researchers to conduct different types of reviews, especially systematic reviews. It is a very important practice for a researcher to set clear research questions for the current research interest and indicate those in a protocol (CRD, 2009). For example, a review about 'selecting dietary supplement ingredients for preserving and protecting the immune system in healthy individuals' (Crawford et al., 2022), was undertaken to answer the following research questions/study objectives:

i) Can certain dietary supplement components effectively maintain and defend the immune system of healthy individuals?
ii) Can specific dietary supplement components effectively maintain and defend the immune system of healthy individuals under stress?
iii) Do people subjected to stressors (such as sleep deprivation, vigorous exercise, or winter conditions) experience faster recovery from acute respiratory tract infections if they've consumed particular dietary supplement ingredients beforehand, compared to those who haven't taken any supplements or received a placebo?

Similar to the study objectives, the inclusion criteria should be indicated in the protocol conducting the review to ensure that the boundaries between research questions/study objectives are clearly defined. It is therefore crucial for researchers/reviewers who are new/beginners to systematic review's methods strategy to be able to understand certain characteristics of PICO(S) and identify key concepts in the study objectives to outline a corresponding inclusion criteria for future steps, including screening. From this, the researchers can avoid common mistakes such as choosing variables for inclusion criteria that do not address the research questions and failing to outline essential variables in the inclusion criteria for assessing the study's results' external validity (Patino and Ferreira, 2018). Furthermore, the inclusion criteria must be broad enough to encompass all relevant studies, or there's a risk of missing important information. However, if they're too broad, it could make it difficult to compare and analyze the data effectively. Additionally, the criteria should be practical to apply, as overly complex or detailed criteria could make the screening process unnecessarily time-consuming (CRD, 2009; Horwitz, 1995; Eysenck, 1994). In summary, it is necessary for researchers who are conducting a systematic review to understand PICO(S) elements and apply those to create inclusion criteria for the research of interest. From there, reviewers can adapt certain elements of the inclusion criteria to the study objectives, which is an important step before moving on to study selection (screening).

METHODOLOGIES IN SCREENING ARTICLES FOR INCLUSION

Effective article screening is a dynamic process that requires thoughtful consideration of the research question, resources, and the desired level of rigor. By delving into these methodologies, researchers can make informed choices that align with the goals of their systematic reviews and meta-analyses. The process of screening articles targets the application of eligibility criteria consistently and indiscriminately to all articles identified in search results, minimizing the risk of errors or bias in evidence synthesis.

Removing Duplicate Reference

Assembling references for systematic reviews involves the use of bibliographic reference management tools, among the commonly used tools in evidence syntheses are Endnote and Eppi Reviewer, often used in conjunction with Microsoft Excel. However, other tools like Mendeley and Abstrackr also find utility in the research community. The choice of a reference management tool depends on the specific needs and preferences of the research team, as these tools continually evolve with varying levels of complexity.

A proficient reference management tool should facilitate the removal of duplicate articles, a crucial step in reducing the overall number of articles for screening. Duplicate articles are common in search results and should be removed where possible before eligibility screening starts (Tramer et al., 1997). Inclusion of duplicates in an evidence synthesis could lead to double counting of data, which might introduce bias, as well as create unnecessary additional screening effort. Many reference management tools enable automated identification and removal of duplicate articles (e.g., 'fuzzy matching' of references in Eppi Reviewer), which may be particularly helpful if large numbers of duplicates are present. However, care should be taken to avoid inadvertently removing articles that are not duplicates. If an automated process is used for identifying duplicates, it should not be assumed that this will always classify the articles accurately. This functionality becomes pivotal when the research team needs to group and analyze articles, ensuring effective progress tracking during eligibility screening and identifying potential disagreements between screeners (Choi et al., 2014).

Several considerations can guide review teams in choosing an appropriate reference management tool. These include accessibility, with some tools like Mendeley being openly accessible, while others like Endnote or Eppi Reviewer may require payment. The tool's capacity for handling references, simultaneous users, and suitability for project management tasks, such as task allocation and progress monitoring, are also factors that researchers may find helpful to consider. As the field of reference management tools is dynamic, research teams need to stay informed about the features and updates of different tools to make informed choices based on their specific requirements.

Manual Screening

Traditional yet foundational, manual screening involves researchers meticulously reviewing each article by hand. This hands-on approach allows for a thorough examination of study characteristics, ensuring alignment with inclusion criteria. We delve into the nuances of this method, exploring its benefits and potential challenges. Articles are usually structured in having a "summarized version" with the title and abstract; and/or a "full text" version. Eligibility screening is applied in a stepwise process to these different levels. Although it is entirely a matter of preference, the recommendation to employ at least two screening filters aligns with best practices in evidence synthesis (Rodriguez et al., 2016).

- First screening (Title/Abstract), by reading the titles and abstracts to remove irrelevant material.
- Second Screening (Full Text), assessment of the full-text version of articles for compliance with eligibility criteria. In this stage, Screeners must provide a list of studies excluded with the reason for exclusion.

The dual-filter approach underscores the significance of both initial relevance checks and a thorough evaluation of the complete article. By doing so, researchers adopt a conservative strategy, ensuring that only articles meeting inclusion criteria progress **(Fig. 1).**

Another approach is to screen titles and abstracts separately. However, it is highly Dependent on the number of articles to screen and the nature of the research question. For example, if a significant number of articles can be excluded from the title alone, then combining the title and abstract screening in a single step may be less efficient. Yet in other cases, insufficient information is available in the title or abstract to make an eligibility decision, or even the abstract may be missing, then the full-text version should be screened for the final decision (CEE, 2013).

Linked Articles and Thorough Decision-Making at the Study Level

Recognizing that multiple articles may stem from a single study, the screening process extends beyond individual articles to include linked articles. Linked articles may arise from different formats, such as conference abstracts, reports, or journal papers, all contributing to a comprehensive understanding of a particular study. Detecting and grouping linked articles ensures that all pertinent data relevant to a study are considered during the eligibility decision-making process (von Elm et al., 2004).

Moreover, the final decision-making occurs at the study level, reinforcing a comprehensive approach to evidence synthesis. This involves grouping linked articles and screening them as a single unit. The meticulous consideration of linked articles adds an extra layer of thoroughness to the screening process, aiming to eliminate duplication and ensure that each study is accurately represented within the synthesis.

Recording Article Screening

A meticulous record-keeping strategy serves as a safeguard against errors, bias, and disputes in the article inclusion process. At each screening step, whether involving titles, abstracts, or full-text examinations, it is imperative to maintain a comprehensive log of decisions made by screeners (Beller et al., 2013; Liberati et al., 2009). Modern reference management tools or relational databases provide efficient means of storing and organizing this information. A well-maintained archive of screening decisions allows for easy retrieval and scrutiny, enabling the research team to justify and defend their study selection if questioned. Consideration should be given to the use of appendices as part of the recording process. Appendices can serve as supplementary documentation, providing additional details on specific aspects of the screening process. For instance, an appendix could include a detailed decision matrix showing the combinations of screener agreements, offering a nuanced perspective on the reliability of the screening decisions. Moreover, the inclusion of an appendix listing studies excluded at the full-text screening step, along with reasons for exclusion, enhances the transparency of the decision-making process (Frampton et al., 2017).

This commitment to thorough documentation, including the use of appendices, not only reinforces the reliability of the evidence synthesis but also facilitates continuous improvement and learning throughout the research endeavor. Appendices offer a structured way to present supplementary information without cluttering the main body of the evidence synthesis report, ensuring that interested readers and stakeholders have access to detailed insights into the screening process.

Figure 1. The screening process for SR/MA

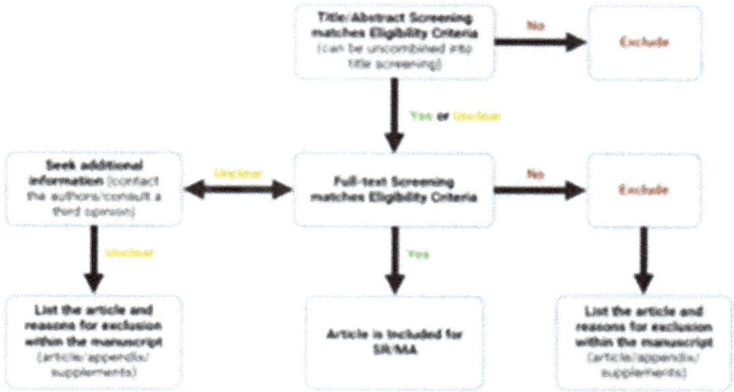

Automated Screening Approaches

The exploration of automated approaches in evidence synthesis refers to the investigation and consideration of technologies such as text mining and machine learning to streamline and enhance the article screening process. Automated approaches aim to leverage computational methods to efficiently identify and categorize relevant articles based on predefined eligibility criteria. These technologies offer the potential to expedite the screening phase, making the process faster and potentially more consistent. Text mining, for example, involves the extraction of valuable information from large datasets, assisting in the initial screening of articles by analyzing titles, abstracts, and even full-text content. Machine learning algorithms can be trained to recognize patterns and criteria, assisting in the classification of articles as relevant or irrelevant (Adeva et al., 2014).

However, the exploration of automated approaches also involves a critical evaluation of their reliability and potential limitations. Researchers need to ensure that these technologies do not introduce new biases or errors into the screening process. Rigorous testing and validation are essential to demonstrate the accuracy and effectiveness of automated methods.

The consideration of automated approaches reflects the ongoing evolution of evidence synthesis methodologies, embracing technological advancements to improve efficiency without compromising the quality and reliability of the screening process.

Screeners and Human Biases

In the context of eligibility screening for research studies, a "screener" is an individual responsible for assessing and determining the eligibility of potential articles or studies based on predefined criteria. The screener plays a critical role in the systematic review or evidence synthesis process by evaluating titles, abstracts, and full-text articles to decide whether they meet the specified inclusion criteria. The screener's task is to filter out irrelevant or ineligible studies, ensuring that the final selection aligns with the objectives and requirements of the systematic review. The screener may be a researcher, subject matter expert, or a member of the review team tasked with making informed and unbiased decisions during the screening phase.

Eligibility decisions are dependent on the reviewer's judgment, so bias or errors may arise if not conducted carefully. An example is to avoid situations where screeners are authors of articles identified in searches by carefully allocating the screening tasks. Another problem that may face the review team is a lack of understanding of information related to eligibility criteria causing wrongful exclusion of articles related to the research question. That's why an appropriate review team should consider the expertise of its members to ensure a reliable screening process (DEFRA, 2015).

An eligibility screening done by one screener is estimated to have an average OF 8% of eligible studies missed, whereas an eligibility screening done by two people working independently would have no studies missed, but having two reviewers to screen eligibility independently increased the number of identified randomized studies by 9% (Edwards et al., 2002). The screeners do not have to be the same two individuals for all articles. Typically, both screeners independently perform the screening process and then compare their decisions. Various methods are available to screen using two individuals. One could be for an individual to screen the articles and the second one to check the first screener's decisions. Yet an independent screening method is preferred as it minimizes the influence of the first screener's decision on the second screener. The main disadvantage in conducting eligibility screening with a single screener, besides selection bias, is that articles can be missed or have errors in their classification and go undetected for various reasons e.g. screener fatigue, human error, etc. What helps in detecting these errors is the double-check process that is created by the presence of a second screener. That explains why many health research systematic reviews consider a mandatory minimum of two screeners (Higgins et al., 2016; IMCSSRCER, 2011).

Resolving Disagreements and Handling Unclear Eligibility Sensitivity Analysis

In the screening process, disagreement between screeners may arise, leading to articles with unclear eligibility status. Resolving any disagreements between screeners should be agreed upon by the review team and pre-specified in the protocol to ensure consistency. To address this, the most efficient approach is for the screeners to meet to discuss their disagreements to reach a decision. Researchers can also adopt a proactive stance, seeking additional information to clarify eligibility where possible. This can be done by attempting to contact the authors of the unclear article. If a decision still can't be reached, then it is advised to counsel a third opinion, from another member of the review team or the project advisory group (Petersen and Ali, 2011). This commitment to transparency and thoroughness ensures that decisions are well-informed and founded on a comprehensive understanding of each article's content. The exact approach may differ; for example, disagreements in the title and abstract screening phase could be discussed by the screeners before handling the full-text screening step so as to avoid unnecessary full-text screening. Another approach is for articles to be directly handled in the full-text screening step so decisions will be based on all available data.

CURRENT CHALLENGES IN THE SCREENING PROCESS

The screening process is a rigorous process which needs precision and accuracy. However, beginner and junior researchers can find it difficult sometimes navigating through the process due to common challenges. In this section, we will go over some common difficulties in the screening process when conducting a systematic review.

A literature search using an established search strategy is an important practice whenever researchers need to conduct systematic reviews. However, the search may yield an extensive amount of articles, especially in some categories of medical research and social sciences, and this may be tiring for researchers when they must select only the relevant studies, which may account for only a small portion to be included in the reviews. Some issues to be considered when conducting the screening steps and potential approaches are illustrated in detail below.

Extensive Amount of Articles

The problem exists due to the large number of articles resulting from the search strategy for interdisciplinary topics in social science and medical research. The broader the topic, the more articles will be retrieved from the search strategy from multiple sources and databases. This is due to the adherence to systematic review guidelines when the reviewers carefully search and assess the relevant articles for review purposes. Usually, a large quantity of articles should be a desirable result if all the articles found are related to the review topics. However, this is usually not the case. For example, according to a PRISMA diagram adapted from a systematic review (Iared et al., 2022), from 1349 articles identified by database searching, the researchers excluded 1114 records and identified those articles as irrelevant studies for review purposes.

Distribution of Human Resources

There are human resources problems when it comes to the screening process with lots of articles resulting from the search. For each systematic review, the Cochrane guidance (Lefebvre et al., 2023) and other guidelines (CRD, 2009; Tawfik et al., 2019) suggest that at least two reviewers should independently conduct the screening process and three reviewers will have the highest accuracy and efficiency, especially including beginner and junior researchers, towards the selection process. This may be efficient for reviews that include only a small to moderate number of studies after the search process but for studies with large amounts of resulting articles, the screening process will be tiring and take a vast amount of time.

Duplication

Due to the large number of articles resulting from the search strategy, duplicated articles may cause problems when individual reviewers attempt to assess the relevant articles for inclusion. Several articles might be released for various reasons, such as translations, presenting results at various follow-up intervals, or reporting different outcomes. However, spotting duplicates can be challenging since they're often concealed (not referenced to each other) and neither authorship nor sample size reliably indicates duplication (CRD, 2009; von Elm et al., 2004). Even one study can have multiple duplications and several reports, and this requires the reviewers to carefully compare multiple publications from one study for any differences or contact the authors for clarification (CRD, 2009). However, this process usually takes lots of time, and the authors may not reply if the studies were conducted a long time ago.

Full-Text Accessibility

Depending on the topics of the reviews, reviewing authors may not be able to access the full-text file of the article for inclusion in the review. This is due to the open-access policy of journals, and multiple journals require monthly or yearly subscriptions for access to their journal articles. In other cases, some articles may be removed due to several reasons, including honest error, non-replicable findings and redundant publication (Wager and Williams, 2011). For access to journals, reviewers can contact other potential researchers for temporary access to the journals and retrieve the full-text files. In other cases, the reviewers can treat those articles as inaccessible to full text with correct labeling on a flow diagram such as the PRISMA flow diagram (Page et al., 2021), or the reviewers can contact the authors for clarification, but the problem persists if the authors do not reply.

BEST PRACTICES FOR RIGOROUS SCREENING

Due to current challenges in the screening process, there is a need for practical guidelines and tips on how to select articles more effectively. In this section, we will summarize current practices from guidelines and software for usage throughout the screening process (Polanin et al., 2019; Waffenschmidt et al., 2023; Harrison et al., 2020). To perform an efficient screening process, we recommend the following steps for the researchers to follow:

Before Screening

- Start by establishing a comprehensive set of standards to encompass a wide variety of articles.
- Gradually adjusting the criteria according to the requirements of the research and its practicality, making sure they are in line with the research question and the specified elements, such as PICO(S) or PEO.
- Customize the criteria to match the goals of the systematic review, taking into account variables like study design, study characteristics, interventions, and outcome measures. Make sure the criteria are sufficiently detailed to answer the research questions, yet feasible in terms of participant recruitment and data collection.
- Provide training on abstract/full-text screening, during which screeners will familiarize themselves with the process and test the

tool by screening a set of 20 to 30 abstracts/full-texts. Repeat this process as needed until the team achieves agreement.

During Title-Abstract/Full-text Screening

- Meet with the screening team on a weekly or every other week basis.
- Thoroughly and uniformly apply the criteria to all identified articles to guarantee unbiased selection and reduce errors or biases.
- Keep detailed documentation of screening determinations, noting reasons for exclusions, to promote openness, consistency, and responsibility.
- Utilize reference manager tools for screening such as EndNote, Mendeley, or Zotero.
- Ensure each abstract/full-text is screened by at least two independent reviewers. We recommend three independent reviewers to achieve better results.
- Resolve any discrepancies that arise during the abstract/full-text screening phase. Motivate screeners by imposing time limits on tasks.

After Screening Ends

- Examine the steps taken and choices made following the completion of the screening process.
- Carefully record the process using a PRISMA flow diagram (Page et al., 2021) or a similar diagram.

SOFTWARES

Identifying softwares for the screening process can be challenging and time-consuming, especially for new/beginner researchers in the medical research field. In medical research, three tools stand out as the most useful and efficient tools in conducting systematic reviews, including EPPI reviewer, Covidence, and Rayyan. EPPI reviewer tools can be used during the screening process for prioritizing the results of study selection and increasing efficiency as they can identify relevant citations earlier in the screening process better than Rayyan tools (Waffenschmidt et al., 2023). Covidence and Rayyan emerge as the most suitable tools to support title and abstract screening for systematic reviews in both the feature analysis and the user survey. Researchers found Covidence and Rayyan easy to use and tend to use those tools again in future research (Harrison et al., 2020). We recommend that beginning and junior researchers should first be familiar with the Covidence and Rayyan tools as they are more user-friendly and easier to use than the EPPI reviewer tool.

CHAPTER SUMMARY AND FUTURE DIRECTION

By accurately defining the characteristics required for article inclusion, researchers enhance the relevance and accuracy of their studies. The transparency afforded by clear inclusion criteria contributes to the reduction of bias and confounding variables, improving internal validity. Moreover, well-defined criteria yield high-quality research outcomes, allowing for accurate screening results when running comparisons between reviewers during a screening phase. Effective communication among research team members is facilitated as these criteria serve as guidance for the whole team to follow. Additionally, clear inclusion criteria enable researchers to perform stratified analyses and consider specific subgroups within the study population, enhancing the depth of data analysis. In summary, the importance of clear inclusion criteria extends across participant selection, data analysis, and interpretation, contributing to the overall quality, and reliability of research studies.

REFERENCES

Adeva, J. J. G., Pikatza, J. M., Ubeda, M., & Ansuategi, E. (2014). Automatic text classification to support systematic reviews in medicine. *Expert Systems with Applications, 41*(4), 1498–1508. DOI: 10.1016/j.eswa.2013.08.047

Beller, E., Glasziou, P., Altman, D. G., Hopewell, S., Bastian, H., Chalmers, I., Gøtzsche, P. C., Lasserson, T. J., & Tovey, D. (2013). PRISMA for Abstracts: Reporting Systematic reviews in journal and conference abstracts. *PLoS Medicine, 10*(4), e1001419. DOI: 10.1371/journal.pmed.1001419 PMID: 23585737

Bettany-Saltikov, J. (2010). Learning how to undertake a systematic review: Part 1. *Nursing Standard, 24*(50), 47–55. DOI: 10.7748/ns.24.50.47.s49 PMID: 20865948

Centre for Reviews and Dissemination. (2009). *Systematic Reviews: CRD's guidance for undertaking reviews in health care*. University of York. https://www.york.ac.uk/inst/crd/guidance.htm

Choi, W., Song, S., Ock, S., Kim, C., Lee, J. B., Chang, W., & Kim, S. (2014). Duplicate publication of articles used in meta-analysis in Korea. *SpringerPlus, 3*(1), 182. DOI: 10.1186/2193-1801-3-182 PMID: 24808996

Collaboration for Environmental Evidence. (2013). *Guidelines for Systematic Review and Evidence Synthesis in Environmental Management*. Version 4.2. Environmental Evidence: www.environmentalevidence.org/Documents/Guidelines/Guidelines4.2.pdf

Crawford, C., Brown, L. M., Costello, R. B., & Deuster, P. A. (2022). Select dietary supplement ingredients for preserving and protecting the immune system in healthy individuals: A systematic review. *Nutrients, 14*(21), 4604. DOI: 10.3390/nu14214604 PMID: 36364865

Department for Environment Food and Rural Affairs. (2015). *Emerging tools and techniques to deliver timely and cost effective evidence reviews*. DEFRA.

Edwards, P., Clarke, M., DiGuiseppi, C., Pratap, S., Roberts, I., & Wentz, R. (2002). Identification of randomized controlled trials in systematic reviews: Accuracy and reliability of screening records. *Statistics in Medicine*, *21*(11), 1635–1640. DOI: 10.1002/sim.1190 PMID: 12111924

Eysenck, H. J. (1994). Systematic Reviews: Meta-analysis and its problems. *BMJ (Clinical Research Ed.)*, *309*(6957), 789–792. DOI: 10.1136/bmj.309.6957.789 PMID: 7950571

Frampton, G. K., Livoreil, B., & Petrokofsky, G. (2017). Eligibility screening in evidence synthesis of environmental management topics. *Environmental Evidence*, *6*(1), 27. DOI: 10.1186/s13750-017-0102-2

Green, S., & Higgins, J. P. T. (2011). *Cochrane Handbook for Systematic Reviews of Interventions*. John Wiley & Sons.

Harrison, H., Griffin, S. J., Kuhn, I., & Usher-Smith, J. A. (2020). Software tools to support title and abstract screening for systematic reviews in healthcare: An evaluation. *BMC Medical Research Methodology*, *20*(1), 7. DOI: 10.1186/s12874-020-0897-3 PMID: 31931747

Higgins, J. P. T., Lasserson, T., Chandler, J., Tovey, D., & Churchill, R. (2016). *Methodological expectations of Cochrane intervention reviews*. Cochrane.

Horwitz, R. I. (1995). "Large-scale randomized evidence: Large, simple trials and overviews of trials": Discussion. A clinician's perspective on meta-analyses. *Journal of Clinical Epidemiology*, *48*(1), 41–44. DOI: 10.1016/0895-4356(94)00171-L PMID: 7853046

Iared, W., Mourão, J. E., Puchnick, A., Soma, F., & Shigueoka, D. C. (2022). Angioplasty versus stenting for subclavian artery stenosis. *The Cochrane Library*, *2022*(2), CD008461. Advance online publication. DOI: 10.1002/14651858.CD008461.pub4 PMID: 35187653

Institute of Medicine (US) Committee on Standards for Systematic Reviews of Comparative Effectiveness Research. (2011). *Finding What Works in Health Care: Standards for Systematic Reviews* (Eden, J., Levit, L., Berg, A., & Morton, S., Eds.). National Academies Press., DOI: 10.17226/13059

Lefebvre, C., Glanville, J., Briscoe, S., Featherstone, R., Littlewood, A., Metzendorf, M.-I., Noel-Storr, A., Paynter, R., Rader, T., Thomas, J., & Wieland, L. S. (2023). *Chapter 4: Searching for and selecting studies*. In: Higgins JPT, Thomas J, Chandler J, Cumpston M, Li T, Page MJ, Welch VA (editors). Cochrane Handbook for Systematic Reviews of Interventions version 6.4 (updated October 2023). Cochrane. https://training.cochrane.org/handbook

Liberati, A., Altman, D. G., Tetzlaff, J., Mulrow, C. D., Gøtzsche, P. C., Ioannidis, J. P. A., Clarke, M., Devereaux, P., Kleijnen, J., & Moher, D. (2009). The PRISMA statement for reporting systematic reviews and meta-analyses of studies that evaluate healthcare interventions: Explanation and elaboration. *BMJ (Clinical Research Ed.)*, *339*(1), b2700. DOI: 10.1136/bmj.b2700 PMID: 19622552

Livoreil, B., Glanville, J., Haddaway, N., Bayliss, H., Bethel, A., De Lachapelle, F. F., Robalino, S., Savilaakso, S., Zhou, W., Petrokofsky, G., & Frampton, G. K. (2017). Systematic searching for environmental evidence using multiple tools and sources. *Environmental Evidence*, *6*(1), 23. DOI: 10.1186/s13750-017-0099-6

Meline, T. J. (2006). Selecting studies for systemic review: Inclusion and exclusion criteria. *Contemporary Issues in Communication Science and Disorders*, *33*(1), 21–27. DOI: 10.1044/cicsd_33_S_21

Page, M. J., McKenzie, J. E., Bossuyt, P. M., Boutron, I., Hoffmann, T., Mulrow, C. D., Shamseer, L., Tetzlaff, J., Akl, E. A., Brennan, S., Chou, R., Glanville, J., Grimshaw, J., Hróbjartsson, A., Lalu, M. M., Li, T., Loder, E., Mayo-Wilson, E., McDonald, S., & Moher, D. (2021). The PRISMA 2020 statement: An updated guideline for reporting systematic reviews. *BMJ (Clinical Research Ed.)*, *372*(1), n71. DOI: 10.1136/bmj.n71 PMID: 33782057

Patino, C. M., & Ferreira, J. C. (2018). Inclusion and exclusion criteria in research studies: Definitions and why they matter. *Jornal Brasileiro de Pneumologia : Publicaçao Oficial da Sociedade Brasileira de Pneumologia e Tisilogia*, *44*(2), 84. DOI: 10.1590/s1806-37562018000000088 PMID: 29791550

Petersen, K., & Ali, N. B. (2011). Identifying Strategies for Study Selection in Systematic Reviews and Maps. In 2011 International Symposium on Empirical Software Engineering and Measurement. 2011 5th International Symposium on Empirical Software Engineering and Measurement (ESEM). IEEE. DOI: 10.1109/ESEM.2011.46

Polanin, J. R., Pigott, T., Espelage, D. L., & Grotpeter, J. K. (2019). Best practice guidelines for abstract screening large-evidence systematic reviews and meta-analyses. *Research Synthesis Methods*, *10*(3), 330–342. DOI: 10.1002/jrsm.1354

Rodríguez, L. G., Hogarth, N. J., Zhou, W., Xie, C., Zhang, K., & Putzel, L. (2016). China's conversion of cropland to forest program: A systematic review of the environmental and socioeconomic effects. *Environmental Evidence*, *5*(1), 21. DOI: 10.1186/s13750-016-0071-x

Tawfik, G. M., Dila, K. S., Mohamed, M. Y. F., Tam, D. N. H., Kien, N. D., Ahmed, A. M., & Huy, N. T. (2019). A step by step guide for conducting a systematic review and meta-analysis with simulation data. *Tropical Medicine and Health*, *47*(1), 46. DOI: 10.1186/s41182-019-0165-6 PMID: 31388330

Tramèr, M. R., Reynolds, D. J., Moore, R. A., & McQuay, H. (1997). Impact of covert duplicate publication on meta-analysis: A case study. *BMJ (Clinical Research Ed.)*, *315*(7109), 635–640. DOI: 10.1136/bmj.315.7109.635 PMID: 9310564

von Elm, E., Poglia, G., Walder, B., & Tramèr, M. R. (2004). Different patterns of duplicate publication. *Journal of the American Medical Association*, *291*(8), 974. DOI: 10.1001/jama.291.8.974 PMID: 14982913

Waffenschmidt, S., Sieben, W., Jakubeit, T., Knelangen, M., Overesch, I., Bühn, S., Pieper, D., Skoetz, N., & Hausner, E. (2023). Increasing the efficiency of study selection for systematic reviews using prioritization tools and a single-screening approach. *Systematic Reviews*, *12*(1), 161. DOI: 10.1186/s13643-023-02334-x PMID: 37705060

Wager, E., & Williams, P. (2011). Why and how do journals retract articles? An analysis of Medline retractions 1988-2008. *Journal of Medical Ethics*, *37*(9), 567–570. DOI: 10.1136/jme.2010.040964 PMID: 21486985

Chapter 8
How a Meta-Analysis Works

Lam D. Chau
https://orcid.org/0000-0002-1215-087X
Case Western Reserve University, USA

Quan K. Nguyen
https://orcid.org/0000-0002-3222-0978
Vin University, Vietnam

ABSTRACT

This chapter serves as a comprehensive guide for researchers, especially junior and beginner researchers in the medical research field. Meta-analysis is explained in detail, emphasizing its role in synthesizing findings from various studies. Basic concepts such as effect sizes, effect models, and heterogeneity assessment are introduced, followed by an overview of the steps involved in conducting and interpreting a meta-analysis. Additionally, the chapter introduces advanced topics including subgroup analysis, meta-regression, and handling complex data structures. Practical recommendations and toolkits from the literature are provided to facilitate efficient meta-analysis. Furthermore, the chapter addresses current limitations and criticisms of meta-analyses, including issues of publication bias, heterogeneity, and subjectivity. By elucidating these challenges and

DOI: 10.4018/979-8-3693-3007-4.ch008

offering potential solutions, the chapter aims to equip researchers with the necessary knowledge to conduct robust meta-analyses and contribute effectively to the field of medical research.

1. INTRODUCTION TO META-ANALYSIS

Meta-analysis refers to the "*analysis of analyses*," i.e., the statistical combination of results from separate studies to produce a single summary result. The term was coined by Gene V. Glass in 1976 as a way to make sense of the rapidly expanding research literature (Greco et al., 2013). In simpler terms, a meta-analysis is like taking multiple pieces of a puzzle [individual study findings] and assembling them to see the bigger picture. This approach offers a rigorous alternative to traditional narrative reviews by using quantitative methods to integrate findings across studies (Borenstein et al., 2021; Higgins et al., 2024). Well-conducted systematic reviews and meta-analyses reside at the top of the evidence hierarchy, generally providing the highest level of evidence for clinical and policy decisions (Al Noman et al., 2024). For example, the American Heart Association assigns its highest evidence level to meta-analyses of multiple trials [Level A], reflecting the confidence warranted by consistent findings across studies (Murad et al., 2016).

1.1. Historical Perspective

While the formal term *meta-analysis* is relatively recent, the idea of combining evidence has existed for over a century. The first known medical meta-analysis was by statistician Karl Pearson in 1904, who pooled data from several studies on typhoid vaccine efficacy (Greco et al., 2013). Pearson reported an aggregate result [and even discussed methodological issues] when assessing enteric fever inoculation in the British Army (Greco et al., 2013; Pearson, 2010). Much later, in the 1970s, Gene Glass introduced *meta-analysis* in social science research (Greco et al., 2013), and scholars began applying it widely in medicine by the 1980s. Pioneers like Archie Cochrane in the 1970s had called for summaries of controlled trials to guide healthcare, paving the way for systematic reviews (Liamputtong et al., 2023).

By 1993, the Cochrane Collaboration was founded to regularly compile evidence from randomized trials, firmly establishing systematic reviews and meta-analyses as key tools in medicine. Since then, the growth of meta-analyses has been exponential. As of April 2025, simple PubMed search found more than 296,000 results with "meta-analysis" in the title or abstract. Of these, 75% were published within the last ten years [since 2016], and nearly half of all-time systematic reviews were published within last five years. This explosion reflects both the need to synthesize overwhelming research information and the recognition that single studies are rarely definitive.

1.2. Why Meta-Analysis?

Individual studies, especially clinical trials, often have limited sample sizes and may report seemingly conflicting results. Meta-analysis increases statistical power by pooling data, which can detect small but important effects that individual studies miss. It also estimates an overall effect size with greater precision [narrower confidence intervals] than any one study. Moreover, by summarizing the totality of evidence, meta-analyses help resolve uncertainty when studies disagree and provide clearer guidance for practice or policy (Al Noman et al., 2024).

For instance, if five trials of a new therapy yield mixed results – two positive, two negative, one inconclusive of effectiveness – a meta-analysis can integrate all their data to determine whether, on balance, the therapy is effective. This cumulative evidence approach is invaluable in medicine, where decisions often cannot wait for a hypothetical "perfect" large trial. Indeed, well-conducted meta-analyses often inform clinical guidelines and health policies with a level of credibility that single studies rarely achieve. At the same time, meta-analysis is not a magic bullet – its validity depends on the quality of included studies and appropriate methodology. As this chapter will explore, *"garbage in, garbage out"* applies: a meta-analysis can only be as reliable as the data it synthesizes. Thus, understanding both the strengths and limitations of meta-analysis is critical for its proper use.

2. PLANNING A META-ANALYSIS: FROM QUESTION TO PROTOCOL

2.1. Research Question With PICO

Every meta-analysis starts with a well-defined *research question.* Typically, this arises from a problem in the literature: for example, "Does treatment X improve survival in condition [or disease] Y?" or "What is the prevalence of outcome Z across studies?" A focused question is often framed in terms of PICO [Population, Intervention, Comparator, Outcome] for clinical interventions (Higgins et al., 2024; Liamputtong et al., 2023). Clarity here guides the entire process, including which studies to include. Early-career researchers should resist overly broad questions; a narrow, clearly scoped question makes the meta-analysis more manageable and the findings more actionable (Greco et al., 2013). For instance:

- Instead of asking: "Do lifestyle interventions help diabetes?",
- One should ask: "In adults with type 2 diabetes, does a low-carbohydrate diet [intervention] lead to greater 6-month glycemic control compared to a low-fat diet [comparator]?".

2.2. Writing and Registering a Protocol: PRISMA and PROSPERO

With a question in hand, the next step is developing a *protocol* (Higgins et al., 2024; Tawfik et al., 2019). A protocol is essentially a recipe for how the review and meta-analysis will be conducted, written *before* looking at study results to minimize bias. It pre-specifies criteria for including studies [e.g. only randomized controlled trials, or only studies with a certain design], the search strategy, outcomes of interest, and planned analyses. Registering the protocol in a database like PROSPERO [an international registry for systematic review protocols] is encouraged for transparency and to avoid duplication of effort (PROSPERO,). Such registration can prevent selective reporting by making public the intended methods and analyses in advance. Think of the protocol as a blueprint – it keeps the meta-analyst on track and enhances the credibility of the final product, since deviations must be reported and justified. Adhering to established guidelines like PRISMA 2020 [Preferred Reporting Items for Systematic Reviews and Meta-Analyses] from the outset can also be helpful (Page et al., 2021). PRISMA provides a 27-item checklist that reminds researchers to report key information such as search dates, risk of bias assessment, and statistical methods.

2.3. Eligibility Criteria

Deciding which studies to include [and exclude] is crucial. Criteria usually cover the population characteristics, study design [e.g. include only RCTs or also observational studies], interventions/exposures, outcomes, and minimum data required. In other words, the core for screening individual studies is also based on PICO question framework [though usually an extended PICODTS might apply] (Liamputtong et al., 2023). It's often a balance between breadth and precision: overly broad inclusion [mixing "apples and oranges"] can introduce heterogeneity that complicates interpretation, whereas too narrow criteria may yield too few studies to be meaningful (Borenstein et al., 2021; Greco et al., 2013; Higgins et al., 2024).

For example, a meta-analysis on telemedicine outcomes might choose to include various telehealth interventions across different conditions – broad but possibly too heterogeneous – or restrict to telemedicine for a specific disease – narrower but more uniform (Dat et al., 2023). Researchers should justify these choices a priori. One common criticism of meta-analysis is the inappropriate combining of diverse studies that really should not be pooled together [later in this chapter and in theme of meta-analysis, this term *pool* will be used frequently]. To avoid this *"apples vs. oranges"* problem, ensure the included studies address fundamentally similar questions. If important differences exist [e.g. one study uses a higher drug dose or a slightly different outcome definition], one might perform subgroup analyses or at least acknowledge the variation during interpretation, rather than lumping everything together blindly and making a hotpot of data.

3. CONDUCTING THE LITERATURE SEARCH

The power of a meta-analysis lies in capturing all relevant studies on the question – not just the ones that are easy to find or those published in high-impact journals. In other words, meta-analysis is nested within a systematic review. In light of this, meta-analyses is nested within a systematic review. Thus, a systematic and exhaustive literature search is essential. Typically, researchers search multiple databases [e.g., PubMed/ MEDLINE, Scopus, Embase, Cochrane Library, Web of Science] using a combination of keywords and subject headings tailored to the question.

For instance, a meta-analysis on breastfeeding during pregnancy would use terms like "breastfeeding," "pregnancy," "pregnant women," and filter for studies that report maternal or fetal outcomes (Minh et al., 2021). The search often includes synonyms and variant terms, with an effort to ensure nothing is missed [in this example, terms like "nursing during pregnancy" or "concurrent lactation and pregnancy" might be included]. It is also important to search grey literature – conference abstracts, theses, trial registries – especially to mitigate publication bias [the tendency for positive results to be published more than negative ones]. Including unpublished or non-traditional sources can uncover studies with null results that would otherwise remain in the "file drawer," skewing the meta-analysis if omitted.

Most systematic reviews document the search strategy in detail [often in an appendix], listing the databases, date of last search, and specific query strings used. Following PRISMA guidelines, researchers then present a flow diagram of study selection – how many records were found, how many screened [after removing duplicates], how many full texts evaluated, and how many finally included, with reasons for exclusions (PRISMA Team,). This transparency allows readers to judge the thoroughness of the search and selection process.

Early-career researchers should learn the art of database searching, possibly consulting a librarian or information specialist to refine strategies. It is common to use Boolean logic [basic AND/OR and advances as needed] and apply inclusion/exclusion filters. The goal is high sensitivity [don't miss any relevant studies] with reasonable specificity [don't retrieve too many irrelevant hits]. One practical tip: after running searches, also scan the reference lists of key articles and prior reviews for any studies the database search might have missed [*snowballing technique*]. This can be surprisingly effective in catching older studies or those in lesser-known journals.

Revisiting the breastfeeding-and-pregnancy example previously mentioned, Minh and colleagues [2021] systematically searched nine databases, yielding 1,992 studies, which were whittled down to 20 eligible studies (Tawfik et al., 2019) (Minh et al., 2021). Such a winnowing process is typical. The PRISMA flowchart in their paper shows the funnel from nearly two thousand titles to the final handful included. This demonstrates both the scope of literature on the topic and the rigorous filtering applied [e.g. many studies might have been irrelevant, duplicates, or of insufficient quality]. The thorough search gave confidence that the meta-analysis in that study truly reflected *all available evidence* on breastfeeding during pregnancy, thereby strengthening the validity of their conclusions.

Once search results are gathered, at least two reviewers usually independently screen titles/abstracts to identify potentially relevant studies, then review full-texts to confirm eligibility. Using two or more reviewers [with a process to resolve disagreements, such as discussion or a third reviewer] helps avoid selection bias and random errors. Modern systematic reviews often use software tools like Covidence or Rayyan to facilitate screening and record reasons for exclusion systematically. It is important to be aware of study selection bias – for example, if one unknowingly includes a disproportionate number of studies from certain regions or excludes small studies due to publication language [e.g., ignoring non-English studies can introduce language bias]. Best practice is to include all languages and use translation if necessary, or justify if language restrictions are applied. Keeping the inclusion process systematic and transparent is crucial. At the end of this stage, you have the final set of studies for data extraction and analysis.

4. DATA EXTRACTION AND MANAGEMENT

4.1. What to Extract?

With the set of included studies in hand, the next phase is extracting data from each study. This means pulling out the relevant information needed for analysis and reporting. Typically, data extraction is done using a standardized form or spreadsheet. Key items to extract include:

- **Study characteristics:** author, year, location, study design, sample size, population details [e.g. age, gender, inclusion criteria].
- **Intervention/exposure details:** what was the treatment or exposure [dose, duration, etc.] and comparator [placebo, standard care, etc.] in each study.
- **Outcomes:** the numerical results for outcomes of interest [e.g. for each group: number of events or event rates for binary outcomes; means and standard deviations for continuous outcomes; hazard ratios, etc., as reported].
- **Measure of variance:** standard errors, 95% confidence intervals, or p-values associated with the outcomes, which are needed to weight studies and compute pooled estimates.
- **Follow-up time:** especially important in clinical studies [e.g. 6-month outcome vs 1-year outcome].
- **Quality metrics:** results of risk-of-bias assessment for each study [e.g. randomization quality, blinding, attrition rates for trials, or Newcastle-Ottawa Scale scores for observational studies].

Data extraction should be performed carefully, ideally by at least two independent reviewers, to minimize errors. One common approach is double data extraction with a consensus process for discrepancies. This redundancy catches mistakes like typographical errors or misreading a figure from a paper. Maintaining a codebook or data dictionary is helpful so that all extractors interpret items consistently [for example, clarifying how to record an outcome reported as median if you are synthesizing means – often such a study might be excluded or a method used to approximate a mean].

4.2. How to Extract?

All data is then entered into a dataset for analysis, commonly using statistical software or specialized meta-analysis software. Tools like RevMan [Review Manager, from Cochrane] provide user-friendly interfaces for entering data and performing meta-analyses (Higgins et al., 2024). Alternatively, software like Excel or CSV can hold the data, which can then be analyzed in R [packages like *metafor*], Stata [the *metan* command], SPSS or others. There are also dedicated programs like Comprehensive Meta-Analysis [CMA], featured by Dr. Borenstein and colleagues, who is an expert and has also been very active in the field for a while (Borenstein et al., 2021). The key is to ensure data accuracy and completeness. It is wise to double-check any extreme or influential values against the source papers.

4.3. What to Do When Data is Missing

During extraction, one may realize that not all studies report data in a readily usable form. Dealing with missing or inconsistent data is a common challenge. For example, one trial might report an odds ratio with confidence interval for an outcome, while another reports raw counts of events, and another only states "no significant difference observed" without details.

The analyst often has to derive comparable metrics. If raw data [like 2x2 tables of events and non-events] can be obtained from each study, that is ideal because one can then compute the effect sizes uniformly. If some information is missing, sometimes authors can be contacted to provide additional data. In Minh et al.'s meta-analysis on breastfeeding during pregnancy, the authors extracted data on various maternal and fetal outcomes [miscarriage rates, preterm birth, hemorrhage incidence, etc.] and noted whether differences between breastfeeding vs not breastfeeding during pregnancy were statistically significant in each study (Minh et al., 2021). They found, after pooling, no significant differences in adverse outcomes between the two groups – a reassurance that was only evident by quantitatively combining the small studies. Extracting those outcome counts and rates from each paper was crucial to compute the combined risk ratios showing equivalence.

4.4. Risk of Bias Assessment

It is also at this stage that you assess study quality and bias. Commonly, systematic reviews will use tools like Cochrane's Risk of Bias tool [for RCTs] or the Newcastle-Ottawa Scale [for observational studies] and extract judgments for each study [e.g. low, unclear, or high risk of bias in various domains] (Sterne et al., 2019; Wells et al.,). Ideally, the tool that be used for risk of assessment should match the design of individual studies. These assessments can later be used in analysis [for example, doing a subgroup meta-analysis of only the high-quality studies to see if results differ]. They also feed into interpreting the overall evidence strength [as in GRADE evaluations – more on that later].

In summary, data extraction is a meticulous but vital process. Well-structured data at this point sets the stage for the statistical synthesis to follow. Early-career researchers should not underestimate the time needed for this step – careful reading of full texts, double-checking numbers, and sometimes creative problem-solving to obtain consistent metrics. Once the dataset is ready, we can proceed to the actual meta-analytical calculations.

5. HANDS ON AND CONDUCTING A META-ANALYSIS

5.1. Choosing Effect Sizes for Meta-Analysis

At the heart of a meta-analysis is the pooling of effect sizes. An *effect size* is a quantitative measure of the outcome or association that each study reports. The choice of effect size depends on the nature of the outcome:

5.1.1. Binary [Dichotomous] Outcomes: Risk Ratio, Odds Ratio, and Risk Difference

For dichotomous outcomes [e.g. event occurred or not, such as death, disease incidence, treatment success], common effect sizes include the risk ratio [RR], odds ratio [OR], or risk difference [RD]. The RR is the ratio of the probability of the event in the treatment group to that in the control group; the OR is the ratio of odds of the event between groups. ORs are often used in meta-analyses of clinical trials because of certain statistical properties, but they can be less intuitive than RRs. An OR can be converted to an approximate RR if needed for interpretation [this applies for longitudinal studies, not cross-sectional]. If outcomes are very rare, OR and RR will be similar. RD [absolute difference in proportions] is less commonly pooled but can be used if an absolute effect measure is desired.

5.1.2. Continuous Outcomes: Mean Difference, and Standardized Mean Differences

For continuous outcomes [e.g. blood pressure level, symptom score], we often use mean difference [MD] if all studies use the same scale, or standardized mean difference [SMD] if studies use different scales measuring the same underlying construct. SMD [often Cohen's d in meta-analysis context] expresses the difference in means in units of the pooled standard deviation. For example, a meta-analysis of pain relief might include some studies using a 0–10 pain scale and others using a 0–100 scale; by standardizing, one can pool these results.

5.1.3. Time-to-Event Outcomes: Hazard Ratio

For time-to-event outcomes [survival analysis], the typical effect size is the hazard ratio [HR] comparing hazard rates between groups, if the included studies report those from Cox models. Alternatively, one can pool survival probabilities at specific time points [treated as a binary event of surviving to X time].

For others, for example, correlation or diagnostic accuracy studies, one might pool correlation coefficients [possibly after Fisher's z-transformation] or measures like sensitivity and specificity. Regardless of the effect size used, the analyst must ensure that all effect sizes are oriented in the same direction [e.g. >1 always favors treatment, or always indicates harm, consistently across studies]. Sometimes that means reversing an effect from a study for consistency. It is also essential to decide whether to use fixed-effect or random-effects models for pooling – a fundamental choice in meta-analysis (Borenstein et al., 2021; Higgins et al., 2024).

5.2. Choosing Effect Models: Fixed- and Random-Effect Models

These are two different statistical models for meta-analysis that reflect how we assume study effects relate to each other:

5.2.1. Fixed-Effect Model

A fixed-effect model assumes there is *one true effect size* [one true underlying effect] which all studies are estimating, and any variability in study results is due solely to chance [sampling error] (Higgins et al., 2024). Under this model, the pooled estimate is an estimate of this common true effect. Fixed-effect meta-analysis gives more weight to larger studies [since they have smaller variance] and effectively ignores between-study variability beyond chance (Higgins et al., 2024). It is appropriate when the studies are functionally identical [e.g. same design, context, populations – a rare scenario] or when the goal is to compute an average assuming a common effect.

5.2.2. Random-Effect Model

A random-effects model assumes that the *true effect size may vary* from study to study [each study has its own true effect, perhaps due to different populations or implementations], and these true effects follow some distribution [usually assumed normal] around an average effect (Borenstein et al., 2021; Higgins et al., 2024). In other words, there is between-study heterogeneity in addition to within-study sampling error.

The random-effects model incorporates an additional variance component – often called τ^2 [tau-squared] – representing the variance of the true effects across studies. The pooled estimate in a random-effects model is the mean of the distribution of true effects [often referred to as the *summary effect*], and study weights are more balanced than in fixed-effect. To be specific, small studies get relatively more weight under random-effects compared to fixed-effect, because the model does not assume a big study's result can completely dominate if heterogeneity exists (Borenstein et al., 2021; Higgins et al., 2024).

In practice, if there is any substantial heterogeneity among study results, a random-effects model is usually more appropriate. It yields a wider confidence interval [reflecting extra uncertainty due to heterogeneity] than a fixed-effect model. If studies are very homogeneous, fixed and random effects will give similar results [τ^2 will be near zero, so random-effects basically converges to fixed-effect]. It is often recommended to use random-effects as a default in medical meta-analyses because it is an conservative approach to allow for potential variability in effects (Borenstein et al., 2021; Higgins et al., 2024). However, if the number of studies is small or if some studies are very small, random-effects estimates can be unstable.

Analysts sometimes check both models. A notable phenomenon: if there is small-study bias [e.g. smaller studies tend to show larger effects], the random-effects pooled effect can be more influenced by those small studies and thus differ from the fixed-effect result. Cochrane's handbook suggests comparing fixed vs random – if they differs, one should explore why (Borenstein et al., 2021; Higgins et al., 2024). Often, a difference indicates presence of heterogeneity or bias [like smaller studies being lower quality or more prone to selective publication].

For example, in a meta-analysis of a new drug, suppose larger trials tended to find only a modest benefit, but several small trials reported huge benefits [perhaps due to publication bias or chance]. A fixed-effect model [dominated by the large trials] might yield a moderate effect, whereas random-effects might lean more towards the larger effect [since it "believes" each small study could be estimating its own higher true effect] and give a higher pooled estimate. In such cases, analysts would be cautious and investigate the heterogeneity and potential bias before trusting the result.

5.3. Statistical Pooling

Once the effect size metric and model are decided, the meta-analyst applies a formula to compute the weighted average. Most often, *inverse-variance weighting* is used – each study's weight is the inverse of the variance of its effect estimate [so more precise studies count more]. In a fixed-effect model, the only source of variance is within-study variance. In a random-effects model, the weight is inverse of [within-study variance + τ^2], where τ^2 is the between-study variance component. The DerSimonian and Laird [1986] method is a classic approach to estimate τ^2 and do random-effects pooling (DerSimonian & Laird, 1986). Modern implementations might use more refined estimators [like restricted maximum likelihood], but the concept is similar.

Nowadays, tools like RevMan, CMA, or R's *metafor* are very helpful in handling these calculations once the researcher input the data. The output will be a summary effect size [e.g. "pooled OR = 1.35, 95% CI 1.10 to 1.65"] along with a confidence interval and typically a p-value for test of effect. However, emphasis should be on the point estimate and CI, not just the p-value. The meta-analysis may also report a chi-square [Q] test for heterogeneity and the I^2 statistic, which we discuss in the next section.

5.4. Assessing Heterogeneity

Heterogeneity refers to the variation in study outcomes beyond what would be expected by chance alone. After pooling studies, one of the first things to examine is whether the studies are consistent or not. Statistical heterogeneity is commonly quantified using the I^2 statistic and tested with the Cochran's Q test (Higgins et al., 2024). Tau-squared [τ^2] is also used, especially for calculating prediction interval [with predicting purposes], and Dr. Borenstein wrote a fantastic guide distincting these statistical nuances (Borenstein et al., 2021).

5.4.1. Cochran's [Q] Test, or Q

This chi-square test checks if the dispersion of study effect estimates is greater than that expected by sampling error. A p-value < 0.10 [some may use 0.05] suggests presence of significant heterogeneity. However, Q has low power when few studies are included and excessive power [detecting trivial heterogeneity] when many studies are included. Thus, it is often not very informative by itself [a non-significant Q does not guarantee homogeneity if sample is small, and a significant Q in a large meta-analysis might not mean an important issue].

5.4.2. I^2 Statistic, or I^2

This is a transformation of Q that describes the percentage of total variability in effect estimates that is due to heterogeneity rather than chance. It is like a semi-quantitative metric to measure heterogeneity. An I^2 of 0% means no observed heterogeneity, and larger values indicate increasing heterogeneity. Rough guidance often referred is: 0–25% as low heterogeneity, ~25–50% as moderate, ~50–75% as substantial, and 75% or higher as considerable heterogeneity. These thresholds are not absolute, but they help interpretation. For example, $I^2 = 50\%$ implies half the variability in study results is due to real differences between studies.

If heterogeneity is high, the meta-analyst should interpret the summary effect with caution and investigate potential reasons for the differences among studies. Potential sources of heterogeneity include differences in patient populations [age, severity of disease], interventions [dose, quality of implementation], outcome definitions, study design, and risk of bias.

5.4.3. Strategies to Address Heterogeneity

Perform subgroup analyses is one way to address heterogeneity. This approach divide studies into subgroups based on a characteristic and pool within each subgroup. For instance, in a meta-analysis of exercise for weight loss, one might separate studies on men vs women, or high-intensity exercise vs moderate, to see if results differ. If heterogeneity drops within subgroups, that characteristic might explain some differences.

Performing meta-regression is an extension of subgroup analysis, where study-level characteristics are treated as covariates in a regression model to see if they statistically explain variation in effect sizes. Meta-regression can handle continuous variables [e.g. mean age of participants, year of publication] or categorical ones [e.g. low vs high risk of bias, location of study] to assess their relationship with the effect size. For example, Thompson and Higgins [2002] demonstrated how meta-regression could be used to explore whether larger treatment effects are seen in studies with shorter follow-up times (Thompson & Higgins, 2002). A caveat: meta-regression is typically underpowered unless there are a substantial number of studies [rule of thumb: at least 10 studies per variable], and it can sometimes find spurious correlations [multiple testing issues]. It should be used exploratory, generating hypotheses rather than proving them (Borenstein et al., 2021).

Use a random-effects model if not already doing so is also a good choice. As discussed, this model incorporates heterogeneity into the analysis, yielding a more conservative [wider CI] summary. If heterogeneity is extreme and cannot be explained, one might even question the wisdom of pooling at all – sometimes a narrative summary might be more appropriate if results are too disparate.

Consider excluding outliers or very heterogeneous studies in a sensitivity analysis can be another option for assessing heterogeneity. See if the pooled result is robust after removing a study that is very different from the others in methodology or results. However, this should be done cautiously and reported transparently [one shouldn't simply drop studies to get a cleaner result without justification, and the results may likely be more prone to bias in that case].

For example, in Minh et al.'s meta-analysis on breastfeeding during pregnancy, the authors found moderate heterogeneity among the included studies for most outcomes (Minh et al., 2021). This consistency across studies may strengthen their conclusion that breastfeeding during a normal, low-risk pregnancy does not adversely affect maternal or fetal outcomes. Conversely, one of the outcome in their Figure 3C shows $I^2 = 75\%$. The researchers might explore differences: perhaps heterogeneity is partly explained by whether trials were industry-funded or not [maybe effect sizes are larger in industry-funded trials]. A meta-regression could test funding source as a predictor of effect size. Or maybe the heterogeneity stems from different doses used across studies – a subgroup analysis might be helpful in this case. Identifying such patterns can provide important insights beyond the overall average effect.

In summary, assessing and addressing heterogeneity is a critical part of meta-analysis. It moves us from simply computing an average effect to understanding *variability* – when and why effects differ. As many articles have been metaphorically putting it, meta-analysts must decide whether they are combining apples with apples, or apples with oranges. If it is the latter, they need to peel back the differences or at least acknowledge them when drawing conclusions.

5.5. The Forest Plot: Displaying Results

Once data are pooled, the results of a meta-analysis are often visualized using a **forest plot**. A forest plot is a graphical representation that shows the effect estimate and confidence interval for each study and the pooled summary estimate. It gets its name perhaps because the set of confidence interval lines can look like a forest of forest trees when there are many studies.

Figure 1. Example of a forest plot

In a typical forest plot [see Figure 1], each study is listed on the left with an identifier [by author and year], and the numerical results [effect size and CI] are often listed to the right. The middle of the plot has a horizontal axis [often logarithmic scale if OR/RR so that a CI that crosses 1

is symmetric; for continuous outcome this null value is 0]. Each study's result is shown as a square [or dot] centered on its effect estimate, with a horizontal line extending on either side to show the conventional 95% confidence interval. The size of the square usually reflects the *weight* of the study in the meta-analysis [larger studies or those given more weight have larger squares]. To remind, this weight is different between fixed- and random-effect models, depending on which method is used for pooling the data. On the same axis, a vertical line indicates the "line of no effect" [e.g. OR = 1, or MD = 0]. If a study's confidence crosses this line, it means *that individual study* did not have a statistically significant effect on its own [at conventional α=0.05].

At the bottom, the diamond represents the pooled effect. The center of the diamond is the summary effect estimate, and its horizontal tips show the confidence interval. If the diamond does not cross the line of no effect, the *pooled result* is statistically significant. Numerical values for the pooled effect and heterogeneity statistics are often printed as well.

Interpreting a forest plot involves both the individual study results and the summary. One quickly sees which studies favored the treatment [e.g. OR < 1 if outcome is bad and treatment reduces it] versus those that favored control [OR > 1], and how consistent the results are. If all study CIs overlap considerably and centers align on one side of no effect, it is a visually consistent picture. If they are all over the place, heterogeneity is apparent. The plot might also display an I^2 value and p for heterogeneity somewhere [usually at the bottom of the figure as a legend], reinforcing what one sees by eye.

Forest plots are very information-dense yet efficient. A quick scan can tell a reader the direction and significance of effects across studies and overall. They also often include a subtotal or overall row, which might be used when subgroups are plotted within the same forest [e.g. separate diamonds for subgroups and an overall diamond]. As a note, ensure the scale is properly noted – for OR/RR, a log scale is typically used in the plot, but the axis ticks might be labeled in the original scale [1, 0.5, 2, etc., spaced logarithmically]. For mean differences, a linear scale is used, and null value is set at 0.

Early-career researchers should practice reading forest plots, as they are ubiquitous in research papers. Key points to check are: [i] Does the *pooled* CI cross no effect or not? [ii] Is there consistency among study results? [iii] Are any studies weighted very heavily [and if so, do they drive the result]? [iv] Are subgroup differences visible? Keep in mind that the visual impression should be backed up by the numerical statistics presented. As an educational aid, Cochrane's Students 4 Best Evidence website provides tutorials on how to read forest plots, which can be useful to solidify these concepts (Students 4 Best Evidence, n.d.).

5.6. Evaluating Publication Bias and More: The Funnel Plot

Even after conducting a rigorous meta-analysis, one must consider the possibility of publication bias – the selective publishing of studies based on their results. Publication bias can skew a meta-analysis if, for example, small studies with null results remain unpublished, while those with flashy positive results are in the literature. One popular tool to detect such bias is the funnel plot, and readers are strongly recommended to visit this page for a visual and interactive learning experience (Students 4 Best Evidence, n.d.).

If the funnel plot is asymmetric – say, missing on one side – it might indicate that certain studies are absent. A common scenario is that, small studies showing no significant effect might be missing [not published], whereas small studies showing large effects are present. This would appear as an asymmetry where the bottom of the funnel has more points on the side of a positive effect, and a gap on the other side. The meta-analyst might suspect that missing studies [the open circles in Figure 2] should have been there to balance the funnel. This asymmetry can result from publication bias, but also from other factors like true heterogeneity correlated with study size [e.g. maybe smaller studies had patients with more severe disease and thus larger treatment effects]. So one must interpret funnel plots cautiously (Choi & Lam, 2016; Sterne & Harbord, 2004).

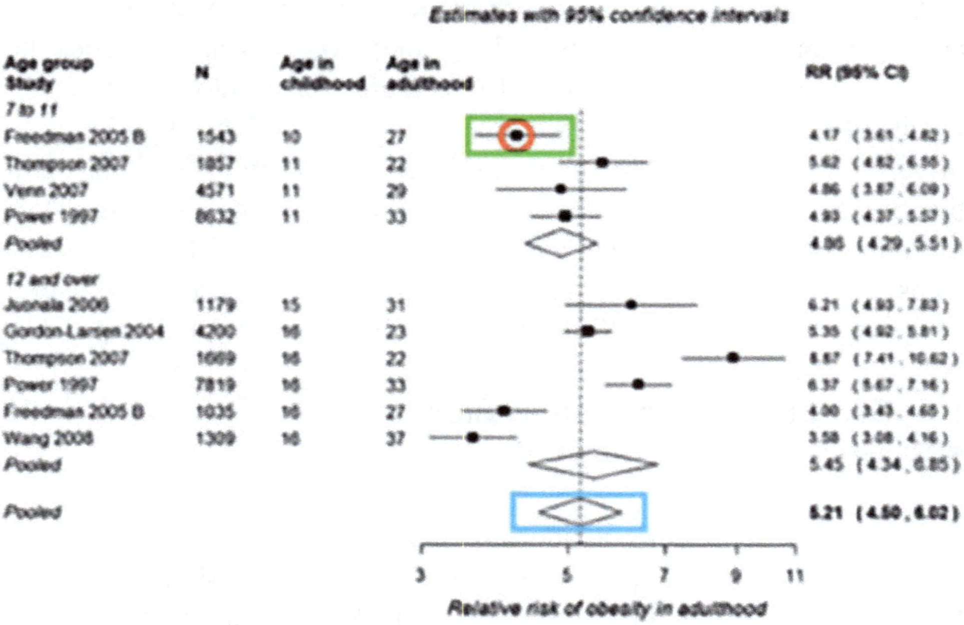

To aid interpretation, formal tests exist, like Egger's test of the intercept [which regresses a standardized effect estimate on precision – a significant intercept suggests asymmetry] (Borenstein et al., 2021; Higgins et al., 2024). Egger's test is sensitive especially when there are a decent number of studies [usually ≥10]. A significant Egger test [$p<0.05$ for bias] provides statistical evidence of asymmetry beyond chance. There's also Begg's test [rank correlation method] and more advanced methods like trim-and-fill [which imputes where missing studies might lie and re-computes an adjusted pooled effect]. No method is perfect, but these analyses can strengthen the case for or against publication bias affecting the results.

Why is publication bias so important? Because it can make treatments look better [or occasionally worse] than they really are. For example, if five small trials found a drug ineffective but weren't published, and five small trials found it effective and were published, a meta-analysis based only on published data would be overly optimistic. Ioannidis [2016] and others have pointed out that the glut of meta-analyses being produced can sometimes be plagued by biases including selective reporting (Ioannidis,

2016). Therefore, a responsible meta-analysis will comment on the possibility of publication bias – often summarizing the funnel plot findings.

In practice, many meta-analyses with ≥10 studies will include a funnel plot in an appendix or figure. As an example, consider a meta-analysis of a certain supplement's effect on reducing cold duration. If small studies with null results weren't published, the funnel might show a gap where those small, null results should be. The researcher might then say: "Visual inspection of the funnel plot suggests asymmetry, and Egger's test was significant [p=0.02], indicating potential publication bias. The true effect might therefore be smaller than the pooled estimate suggests." They might even apply a trim-and-fill method which could insert hypothetical studies to estimate how much the effect size might drop if the missing studies were included, and report that as a sensitivity analysis.

For early-career researchers, it's important to understand that absence of evidence [no studies] is not evidence of absence [no effect]. Publication bias is a reminder of this adage. Always consider whether the evidence available might be a biased subset of all studies that have been done. Pre-registration of trials and results reporting [like via ClinicalTrials.gov] is helping to reduce this bias, but it remains an ever-present concern in meta-analysis. Thus, checking a funnel plot is a routine part of a meta-analyst's toolbox for quality control.

6. PRACTICAL TIPS FOR EARLY-CAREER RESEARCHERS

Embarking on a first meta-analysis can be daunting. Here are some practical tips and resources:

- **Learn by example:** Read several well-conducted systematic reviews/meta-analyses in your field. Focus on how authors formulate their question, present their methods, and discuss their findings. Cochrane Reviews are often a gold standard in terms of rigor and reporting. Non-Cochrane journals also publish excellent reviews – look for ones with clear methodology sections and insightful discussion of limitations.

- **Use reporting checklists:** PRISMA 2020 is essential for writing up your meta-analysis. For meta-analyses of observational studies, consider the MOOSE guidelines [Meta-analysis Of Observational Studies in Epidemiology] (Brooke et al., 2021). Adhering to these will ensure you don't forget key items like describing bias assessment or heterogeneity exploration. Again, transparency is key, and it really affects the overall quality of the article.
- **Software and tools:** If you're just starting, *RevMan* [free from Cochrane] is a user-friendly way to do basic meta-analysis with an interface to enter data and automatically generate forest plots and stats. *R* [programming] with the *metafor* or *meta* package is more powerful and flexible, especially for meta-regression or advanced plots, but has a learning curve in coding. *Stata* is very popular among epidemiologists for meta-analysis, with commands like *metan* or *metareg*, and many user-written add-ons. *Comprehensive Meta-Analysis [CMA]* is a point-and-click commercial software tailored for meta-analysis that is very friendly to new user, which is worth considering. Whatever you choose, do some test runs with known datasets [some textbooks or courses provide practice data] to be sure you know how to input data correctly [e.g., does the software expect log ORs or raw ORs, etc.?].
- **Quality assessment:** Always assess the risk of bias of included studies. Use appropriate tools that match study design [Cochrane RoB 2 for trials, ROBINS-I for non-randomized, Newcastle-Ottawa for observational, etc.]. Present this in a table or figure. It informs how much you trust the evidence. If many studies are high risk of bias, your conclusion might be tempered ["evidence suggests X, but confidence is limited due to risk of bias"]. Using GRADE [Grading of Recommendations Assessment, Development and Evaluation] is a way to rate the certainty of evidence across studies [high, moderate, low, very low]. While GRADE involves some subjectivity, it systematically considers limitations like risk of bias, inconsistency [heterogeneity], indirectness, imprecision, and publication bias. It's good to at least be aware of GRADE; many health meta-analyses use it to communicate how much confidence we have in the pooled estimate.

- **Do not ignore small details:** Check that studies indeed measure the same construct in outcomes. Verify whether an OR or RR was adjusted for confounders [especially in observational studies] – mixing adjusted and unadjusted estimates in a meta-analysis can be tricky [often one uses adjusted if consistently available]. Also, double-check any computational output. It's easy to accidentally enter an event count in the wrong column and flip an effect; a quick sense-check [e.g., does the pooled result roughly align with a weighted average you might eyeball?] is useful.
- **Dealing with complications:** You might encounter studies with multiple treatment groups or multiple outcomes. For multiple groups [e.g. a trial compares Drug A, Drug B, and placebo], you must avoid double-counting the placebo group if you include two comparisons. One solution is to split the shared group's sample size across comparisons, or choose one pairwise comparison per meta-analysis. For multiple outcomes, keep each meta-analysis focused [don't mix different outcomes in one forest plot; do separate analyses or perhaps a composite if justified]. If studies report outcomes at different time points, you may have to decide one time point of focus or perform separate analyses by time [e.g. 6-month vs 12-month results].
- **Consult a statistician or seasoned reviewer:** Especially for first-timers, having someone experienced to discuss the plan and trouble-shoot is invaluable. Many universities have epidemiologists or bio-statisticians who collaborate on systematic reviews. They can help ensure the model assumptions are correct, and help with advanced methods if needed.
- **Time management:** Doing a meta-analysis is time-consuming. Screening thousands of titles, reading potentially hundreds of papers, extracting data – it can take months. Plan accordingly and work systematically. Maintain good records [e.g., use a reference manager to keep track of all search results, and tag which were included/excluded]. This will save headaches later if you need to justify why a certain study was excluded or update the search.
- **Stay up to date:** If your review takes a long time, new studies might emerge before you finish. Ideally, do a quick update search before finalizing the manuscript to see if any new study has been published

that fits your criteria, and include it if so [or at least acknowledge it]. It's frustrating but not uncommon to see "a new trial came out while our review was in revision; including it did not change the conclusions" – it's good to be thorough up to the last moment.

- **Interpretation and presentation:** Be honest about limitations. If heterogeneity is high, say so and suggest caution – perhaps offer hypotheses for why. If evidence is largely from small or low-quality studies, one might say "the evidence suggests X but further large high-quality trials are needed to confirm findings." Avoid language that overstates certainty. Also, discuss any **external context**: are the results consistent with prior reviews? Did any big trials appear after the included studies that could change things? How does this fit with biological or clinical expectations? For example, if meta-analysis finds a drug effective but a more recent big trial was actually negative, that discrepancy should be discussed. Or if a meta-regression found dose-dependent effects, mention what that implies [maybe only high doses are truly effective in practice].
- **Tables and figures:** In addition to forest plots, many meta-analyses include a table of included studies [summarizing their characteristics] and sometimes a table of results by subgroup. Use these to help readers navigate. Clear visuals [forest and funnel plots] with proper labels as we described make it easier for readers to grasp your findings quickly.
- **Reference resources:** Textbooks like *"Introduction to Meta-Analysis" by* (Borenstein et al., 2021), or the *Cochrane Handbook* (Higgins et al., 2024) are excellent detailed guides. The Cochrane Handbook in particular is freely available online and covers practically every scenario with examples, and it is updated regularly. It also provides rationales and the literature behind methods, which can be useful when writing your methods section and justifying choices.

7. CONCLUSION AND FUTURE DIRECTIONS

Meta-analysis, as part of systematic review methodology, has revolutionized how we summarize and utilize evidence. Instead of relying on one or two influential studies or the loudest expert opinion, we now have a framework to look at *all* the evidence together and come to more objective conclusions. This chapter covered the journey of a meta-analysis: from formulating a clear question, systematically gathering studies, extracting and pooling data, to interpreting and validating the findings with an eye on heterogeneity and bias.

For early-career researchers, understanding meta-analysis is imperative. Even if you are not conducting one, the ability to critically read one is crucial in the era of evidence-based medicine. Clinical guidelines, for instance, heavily depend on meta-analyses to formulate recommendations [in evidence tables, you'll often see references to pooled effect estimates]. Public health policies on screening, vaccination, or disease prevention are often justified by systematic reviews of the literature. As such, meta-analysis acts as a bridge between research and practice – it is the mechanism by which individual studies [the bricks] are assembled into a solid wall of knowledge.

Moreover, contributing to evidence synthesis early in your career can be very rewarding. It nudges you to thoroughly learn a body of literature and to think deeply about study design and bias. It also often reveals where evidence is lacking, spurring new research questions. Many meta-analyses conclude with "more studies are needed, particularly in X subgroup or using Y outcome," thereby shaping the research agenda. In our examples, the telepharmacy review identified a need for more rigorous and quantitative outcome evaluations in telepharmacy, and the breastfeeding meta-analysis highlighted the lack of data in high-risk pregnancies – in each case pointing out areas for future inquiry.

The value of evidence synthesis is perhaps best exemplified by collaborative global efforts: organizations like the Cochrane Collaboration and the Campbell Collaboration [for social interventions] maintain thousands of up-to-date systematic reviews that inform everything from WHO guidelines to local hospital protocols. During the COVID-19 pandemic, rapid meta-analyses helped make sense of conflicting treatment results and guided practice [e.g., confirming dexamethasone's benefit in severe COVID by pooling trials]. This underscores how meta-analysis is not an academic exercise in abstraction, but a pragmatic tool that can save lives by getting the evidence right.

In closing, while one should always be mindful of the limitations, a well-conducted meta-analysis provides a high level of evidence by leveraging the power of many. It embodies a scientific ethos of cumulative knowledge – recognizing that progress comes from synthesizing findings across studies, not from any single study in isolation. Early-career researchers who embrace this will not only contribute to better decision-making in medicine and public health, but also gain a deeper appreciation for rigorous research and critical thinking. As we move forward, the integration of traditional meta-analytic techniques with new data sources and technologies [like real-world data and AI] promises to further enhance evidence synthesis. But the core principles will likely remain: clarity, rigor, transparency, and critical evaluation. Armed with these, researchers are well on their way to understanding and conducting impactful meta-analyses – an essential skill in the evidence-based era.

REFERENCES

Al Noman, A., Sarkar, O., Mita, T. M., Siddika, K., & Afrose, F. (2024). Simplifying the concept of level of evidence in lay language for all aspects of learners: A brief review. *Intelligent Pharmacy*, *2*(2), 270–273. DOI: 10.1016/j.ipha.2023.11.002

Borenstein, M., Hedges, L. V., Higgins, J. P. T., & Rothstein, H. R. (2021). *Introduction to meta-analysis*. John Wiley & Sons. DOI: 10.1002/9781119558378

Brooke, B. S., Schwartz, T. A., & Pawlik, T. M. (2021). MOOSE reporting guidelines for meta-analyses of observational studies. *JAMA Surgery*, *156*(8), 787–788. DOI: 10.1001/jamasurg.2021.0522 PMID: 33825847

Choi, S. W., & Lam, D. M. H. (2016). Funnels for publication bias—Have we lost the plot? *Anaesthesia*, *71*(3), 338–341. DOI: 10.1111/anae.13355 PMID: 26693674

Dat, T. V., Tu, V. L., Quan, N. K., Minh, N. H., Trung, T. D., & Le, T. N.. (2023). Telepharmacy: A systematic review of field application, benefits, limitations, and applicability during the COVID-19 pandemic. *Telemedicine Journal and e-Health*, *29*(2), 209–221. DOI: 10.1089/tmj.2022.0043 PMID: 35758766

DerSimonian, R., & Laird, N. (1986). Meta-analysis in clinical trials. *Controlled Clinical Trials*, *7*(3), 177–188. DOI: 10.1016/0197-2456(86)90046-2 PMID: 3802833

Greco, T., Zangrillo, A., Biondi-Zoccai, G., & Landoni, G. (2013). Meta-analysis: Pitfalls and hints. *Heart, Lung and Vessels*, *5*(4), 219–225. PMID: 24364016

Higgins, J. P. T., Thomas, J., Chandler, J., Cumpston, M., Li, T., Page, M. J., ... (2024). *Cochrane handbook for systematic reviews of interventions* (Version 6.5, updated August 2024). Cochrane. https://training.cochrane.org/handbook

Ioannidis, J. P. A. (2016). The mass production of redundant, misleading, and conflicted systematic reviews and meta-analyses. *The Milbank Quarterly, 94*(3), 485–514. DOI: 10.1111/1468-0009.12210 PMID: 27620683

Khoi Quan, N., & Liamputtong, P. (2023). Systematic review and evidence synthesis in public health. In Liamputtong, P. (Ed.), *Handbook of social sciences and global public health* (pp. 1–25). Springer., DOI: 10.1007/978-3-030-96778-9_143-1

Minh, L. H. N., Tawfik, G. M., Ghozy, S., Hashan, M. R., Nam, N. H., Linh, L. K., Abdelrahman, S. A. M., Quynh, T. T. H., Khoi Quan, N., Nhat Le, T., Ibrahim, H. Y., El-Nile, M. O., Kamel, A. M. A., Giang, H. T. N., & Huy, N. T. (2021). Feto-maternal outcomes of breastfeeding during pregnancy: A systematic review and meta-analysis. *Journal of Tropical Pediatrics, 67*(6), fmab097. Advance online publication. DOI: 10.1093/tropej/fmab097 PMID: 34962568

Murad, M. H., Asi, N., Alsawas, M., & Alahdab, F. (2016). New evidence pyramid. *Evidence-Based Medicine, 21*(4), 125–127. DOI: 10.1136/ebmed-2016-110401 PMID: 27339128

Page, M. J., McKenzie, J. E., Bossuyt, P. M., Boutron, I., Hoffmann, T. C., Mulrow, C. D., Shamseer, L., Tetzlaff, J. M., Akl, E. A., Brennan, S. E., Chou, R., Glanville, J., Grimshaw, J. M., Hróbjartsson, A., Lalu, M. M., Li, T., Loder, E. W., Mayo-Wilson, E., McDonald, S., & Moher, D. (2021). The PRISMA 2020 statement: An updated guideline for reporting systematic reviews. *BMJ (Clinical Research Ed.), 372*(71), n71. Advance online publication. DOI: 10.1136/bmj.n71 PMID: 33782057

Pearson, K. (2010). Report on certain enteric fever inoculation statistics (Original work published 1904). *BMJ*. Republished in *The James Lind Library*. https://www.jameslindlibrary.org/pearson-k-1904/

PRISMA Team. (n.d.). *PRISMA flow diagram generator (Shiny app)*. https://estech.shinyapps.io/prisma_flowdiagram/

PROSPERO. (n.d.). *International prospective register of systematic reviews*. Centre for Reviews and Dissemination, University of York. https://www.crd.york.ac.uk/prospero/

Sterne, J. A. C., & Harbord, R. M. (2004). Funnel plots in meta-analysis. *The Stata Journal*, *4*(2), 127–141. DOI: 10.1177/1536867X0400400204

Sterne, J. A. C., Savović, J., Page, M. J., Elbers, R. G., Blencowe, N. S., Boutron, I., Cates, C. J., Cheng, H.-Y., Corbett, M. S., Eldridge, S. M., Emberson, J. R., Hernán, M. A., Hopewell, S., Hróbjartsson, A., Junqueira, D. R., Jüni, P., Kirkham, J. J., Lasserson, T., Li, T., & Higgins, J. P. T. (2019). RoB 2: A revised tool for assessing risk of bias in randomised trials. *BMJ (Clinical Research Ed.)*, *366*, l4898. DOI: 10.1136/bmj.l4898 PMID: 31462531

Students 4 Best Evidence. (n.d.). *Home*. https://s4be.cochrane.org/

Tawfik, G. M., Dila, K. A. S., Mohamed, M. Y. F., Tam, D. N. H., Kien, N. D., Ahmed, A. M., & Huy, N. T. (2019). A step-by-step guide for conducting a systematic review and meta-analysis with simulation data. *Tropical Medicine and Health*, *47*(1), 46. Advance online publication. DOI: 10.1186/s41182-019-0165-6 PMID: 31388330

Thompson, S. G., & Higgins, J. P. T. (2002). How should meta-regression analyses be undertaken and interpreted? *Statistics in Medicine*, *21*(11), 1559–1573. DOI: 10.1002/sim.1187 PMID: 12111920

Wells, G. A., Shea, B., O'Connell, D., Peterson, J., Welch, V., Losos, M., ... (n.d.). *The Newcastle–Ottawa Scale (NOS) for assessing the quality of nonrandomised studies in meta-analyses*. https://www.ohri.ca/programs/clinical_epidemiology/oxford.asp

Chapter 9
Heterogeneity and Prediction Intervals in Meta-Analysis

Nguyen Khoi Quan
https://orcid.org/0000-0002-3222-0978
Bloomberg School of Public Health, Johns Hopkins University, USA

ABSTRACT

This chapter guides understanding and interpreting heterogeneity [within- and between-study variation] in meta-analyses, crucial for applying findings. We explain why heterogeneity matters and detail four key metrics: Cochran's Q, I^2, τ^2, and H statistic. For each, the chapter will clarify meaning, strengths, limitations, and interpretation, stressing they offer complementary views and shouldn't be used in isolation. Prediction intervals are introduced as a tool to translate statistical heterogeneity into clinically meaningful forecasts of future study effects. We provide guidance on addressing high heterogeneity via subgroup analyses, meta-regression, and outlier/sensitivity analyses, emphasizing transparent reporting [PRISMA 2020]. This practical guide promotes a nuanced understanding of heterogeneity's statistical and clinical implications in meta-analysis, moving beyond simplistic metric interpretation.

DOI: 10.4018/979-8-3693-3007-4.ch009

INTRODUCTION

Heterogeneity – the variation in study results – lies at the heart of interpreting meta-analyses. Even after choosing a fixed-effect or random-effects model, meaningful differences between studies affect conclusions. As mentioned in previous chapters, metrics like Cochran's Q [a chi-square test for heterogeneity], the I^2 statistic, the between-study variance [τ^2], and the derived *prediction interval* [PI] help quantify and interpret heterogeneity in meta-analyses. This chapter explains these concepts in plain language, with practical guidance for early-stage analysts. The chapter will emphasize why heterogeneity matters for evidence synthesis, how to interpret Cochran's Q, I^2, τ^2, and prediction intervals, and what to do when heterogeneity is high. Throughout, we use examples from recent literature to illustrate pitfalls and best practices.

1. WHY HETEROGENEITY MATTERS

Meta-analysis combines results from multiple studies to estimate an average effect. This makes sense only if studies are sufficiently similar [i.e. combinable]. Heterogeneity refers to the variability in true effects or outcomes across studies. It arises from clinical differences [e.g. patient populations, interventions, settings] or methodological differences [e.g. study design, quality]. Even when a random-effects model is chosen to allow differences, substantial heterogeneity means that studies are *not all estimating the same effect*, and the average may be less generalizable than it appears.

For example, a recent systematic review of telepharmacy interventions during COVID-19 found that remote counseling and dispensing helped address pharmacist shortages and improve patient care (Dat et al., 2023). However, the types of telepharmacy models varied widely. Such variability [clinical and operational heterogeneity] means one should be cautious about pooling all results into a single summary. Likewise, another meta-analysis of breastfeeding during pregnancy reported no significant harm (Minh et al., 2021). But if those studies differed in population risk or health systems, the pooled "no difference" could mask important subgroups.

In public health evidence synthesis, it is worth noting that heterogeneity of included research is a major challenge. Awareness of heterogeneity is critical for honest interpretation: high heterogeneity may suggest giving more weight to ranges of plausible effects rather than a single average (Higgins et al., 2019; Ioannidis et al., 2007).

Even after choosing a model, heterogeneity matters. In a fixed-effect model, we assume all studies share one true effect; evidence of heterogeneity would invalidate that assumption. In a random-effects model, we allow different true effects, but heterogeneity still signals uncertainty. Wide variation among study results means that the average effect may not predict outcomes well in any single setting. For example, Borenstein (2023) stresses that I^2 [explained below] "does not tell us how much the effect size varies [except when I^2 is zero]" (Borenstein, 2023). Thus, tools like prediction intervals are essential to convey the range of effects researchers might expect.

2. STATISTICAL METRICS OF HETEROGENEITY

Heterogeneity can be quantified by many different statistics. This section will briefly introduce some common metrics, explaining what they are and how to interpret them.

2.1. Cochran's Q

This is derived from a chi-square test that assesses whether observed variation across study effect estimates is greater than expected by chance. Under the null hypothesis [H_0: no heterogeneity], $Q \approx \chi^2$ with $k - 1$ degrees of freedom [where k is number of studies].

A significant Q [small p-value] suggests heterogeneity. However, Q has limitations: with few studies it has low power [often fails to detect real heterogeneity], whereas with many studies it can be over-sensitive (Borenstein, 2023; Borenstein et al., 2021; Ioannidis et al., 2007). For example, Q may be non-significant simply because only 3 small trials are combined, even if effect sizes differ. Conversely, with 50 trials even trivial differences can make Q "significant." Thus, Q is best viewed qualitatively, and its p-value should be interpreted alongside other information.

2.2. I^2 [I-squared]

This is the most common heterogeneity metric, representing the percentage of total variation in study estimates that is due to true heterogeneity rather than chance. I^2 takes on values range from 0% [no observed heterogeneity] to 100% [all variation due to differences in true effects].

For example, $I^2 = 0\%$ indicates that observed differences are compatible with sampling error; $I^2 = 50\%$ means half the variability is beyond chance, similarly for other thresholds at 25% and 75% (Higgins & Thompson, 2002). However, these cutoffs are context-dependent [small studies often give higher I^2 even with modest differences]. Importantly, I^2 is a relative measure: it depends on how precisely each study estimates its effect. Many authors have noted that I^2 does *not* convey the magnitude of effect differences (Borenstein, 2023; Higgins & Thompson, 2002). Two scenarios can have the same I^2 but very different clinical implications.

2.3. τ^2 [Tau-squared]

The estimated variance of the true effect sizes under a random-effects model. In other words, imagine the true effects [if each study had infinite precision] are distributed normally with variance τ^2. A larger τ^2 means more spread among true effects.

Unlike I^2, τ^2 is on the scale of the effect measure squared [e.g. if effects are log-odds ratios, τ^2 is in squared log-OR units]. There is no universal "high" τ^2 because it depends on the outcome measure. For instance, $\tau^2 = 0.04$ in log-OR corresponds to a standard deviation [τ] of about 0.2, which may or may not be clinically important.

Interpreting τ^2 often involves thinking in terms of the prediction interval [see next section]. When τ^2 is 0, it suggests all true effects are [statistically] the same, but note that as Borenstein points out, in practice "no two studies are truly identical," so $\tau^2=0$ is usually an underestimate unless multiple large trials truly agree (Borenstein, 2023; Higgins & Thompson, 2002).

2.4. H Statistic

Defined as $H^2 = Q/(k - 1)$ [with H is the square root of that ratio], the H statistic measures how much more variation there is in the results than is expected by chance. Think of it this way: if we divide a measure of the total variation [called Q] by the number of studies we looked at [its degrees of freedom], the result helps us understand how different the study results really are (Higgins & Thompson, 2002). It represents the factor by which the observed variance is larger than expected under homogeneity. $H^2 = 1$ means no heterogeneity; $H^2 > 1$ indicates excess variance.

Unlike I^2 statistic, the H^2 is *not* a percentage. It has seen less use than I^2, but one advantage is that H^2 does not truncate at 100%. It offers an alternative perspective: e.g., $H^2 = 1.5$ [meaning the total variance is 1.5 times the within-study variance] corresponds to $I^2 \approx 33\%$. If H^2 is big [say >2], heterogeneity is clearly substantial.

2.5. Put It All Together

Cochran's *Q* provides a formal starting point for judging heterogeneity, but in practice its p-value should be treated as a "smoke alarm," not a verdict. Imagine you have only three trials of a new antiviral: even if their point estimates range from a risk ratio [RR] of 0.5 to 1.8, *Q* may be non-significant simply because the df = 2 test is under-powered. The wise analyst therefore reports *Q* but immediately moves on to effect-size plots and other metrics.

Conversely, in a network meta-analysis of 60 head-to-head comparisons, trivial quirks can drive *Q* to $p < 0.001$. A careful analyst should ask, "Does this statistical flag reflect anything clinically relevant?" A quick rule of thumb is to view *Q* as qualitative evidence – if it is clearly non-significant with many precise studies, genuine consistency is plausible; if it is significant, assess I^2, τ^2 and the prediction interval before deciding how concerning that inconsistency really is.

I^2 translates the chi-square signal into an intuitive percentage: "How much of what I am seeing is beyond chance?" Yet who is new to this often forget that I^2 is scale-free: a meta-analysis of very small trials can yield I^2

$\approx$ 70% even when the absolute spread is narrow, whereas ten mega-trials differing by only a hair may still register $I^2 \approx 5\%$. Therefore, always ask two follow-up questions: (Dat et al., 2023) *How precise are the studies?* and (Minh et al., 2021) *Does the outcome metric matter* (Borenstein et al., 2021; Higgins et al., 2019)? For log-odds ratios, a seemingly modest τ^2 of 0.04 [$\tau \approx 0.20$] corresponds to an odds ratio SD of about 22%; for risk differences measured in percentage points it could be trivial. A practical habit is to calculate – and report – the 95% CI for I^2 itself [most packages provide it], because an I^2 of 45% with a CI of 0% to 78% screams "uncertain heterogeneity," not "moderate heterogeneity".

τ^2 is the real workhorse when one wants to translate heterogeneity into expected outcomes. Begin by choosing an estimator that fits the data [restricted maximum likelihood, Paule–Mandel, or DerSimonian–Laird if the number of studies is small and within-study variances are homogeneous]. Next, convert τ to the original effect scale so clinicians can "feel" it: when $\tau = 0.30$ on the log-hazard scale, tell them it means the true hazards vary roughly $\pm$ 35% around the mean [because $e^{0.3} \approx 1.35$] (Higgins et al., 2019; Higgins & Thompson, 2002). Then show the 95% prediction interval: "Based on $\tau^2 = 0.09$, a hospital starting telepharmacy could see readmission RRs anywhere from 0.62 to 1.16." This single sentence links statistics directly to bedside applications (Borenstein, 2023; Guddat et al., 2012; IntHout et al., 2016).

H^2 sounds arcane, but it is useful in two situations. First, when I^2 is near 100% [common with sparse binary data], H^2 continues to grow and thus discriminates between "lots" and "tons" of heterogeneity. A meta-analysis might report $I^2 = 97\%$ – scary, but not as informative as noting $H^2 = 32$, meaning total variability is 32-fold higher than sampling error. Second, H^2 avoids the perception that heterogeneity has a maximum; readers intuitively grasp that variance can balloon indefinitely (Borenstein et al., 2021; Higgins & Thompson, 2002).

In daily practice, it is good practice to interpret I^2 and τ^2 *together*, never in isolation. If I^2 is 80% but τ^2 translates to only a $\pm$10% variation in risk, the meta-analysis may still be useful for policy makers. Conversely, I^2 of 25% with τ^2 implying a twofold swing in odds is a red flag. Pair those numbers with the prediction interval: a pooled RR of 0.85 [95% CI:

0.76–0.95] but a PI of 0.55 – 1.32 tells guideline developers that benefit is probable yet far from guaranteed in every setting.

By putting these interpretations altogether into your research, you move beyond "statistical housekeeping" and give your audience actionable insight into how, and how much, study results actually vary.

3. USING PREDICTION INTERVALS FOR PLANNING AND GUIDELINES

This concept, which may seem new, often confuses researchers who are new to the field. Prediction intervals [PIs] are the natural companion to random-effects meta-analysis because they acknowledge what clinicians really want to know: "where might the true effect of the next, similar study land?" (Higgins et al., 2024). Whereas the 95% confidence interval [CI] summarises uncertainty around the *average* pooled effect, the 95% prediction interval *predicts* the full range of plausible study-specific effects by combining that average with the between-study variance [τ^2]. As soon as τ^2 is appreciable, the PI is inevitably wider than the CI and can even straddle the line of no effect despite a formally "significant" pooled result – an apparent contradiction that simply reflects heterogeneity, not statistical error (Borenstein et al., 2021).

3.1. Overview

To put in a one-sentence definition, a 95% PI is the interval within which we expect the effect estimate of a *new, comparable study* to fall 95% of the time, given the heterogeneity already observed (Higgins et al., 2024; Riley et al., 2011). This wording deliberately embeds two safeguards for early-career readers: the scope ["comparable study"] and the conditional nature of the probability.

PIs are meaningful only under a random-effects model; the fixed-effect framework assumes one common truth and therefore offers nothing to predict (Borenstein et al., 2021; Higgins et al., 2019). Even under random effects, PIs become unstable when only a handful of studies inform τ^2. A pragmatic rule of thumb – endorsed in the Cochrane Handbook – is to treat PIs based on fewer than five studies as exploratory signals rather than decision-makers (Higgins et al., 2024; IntHout et al., 2016).

In order to compute a 95% PI, researchers who are new to the field need to remember just three inputs: the pooled effect [$\hat{\theta}$], its standard error, and τ^2. Feed those numbers to a software for meta analysis [e.g., the *metafor* package in R [v 4.2-0, default REML estimator], or the *RevMan* 6.5 calculator], and the PI will pop out alongside the CI. For this process, appropriately citing the tool and version protects transparency without forcing readers to reproduce the delta method in algebra (Riley et al., 2011).

In terms of interpretation, two quick questions capture most of the story are about the *direction* and the *width* of the PI and the CI.

1. **Direction** – Does the PI cross the null? If yes, benefit cannot be guaranteed and harm cannot be ruled out.
2. **Width** – How much broader is the PI than the CI? The larger the gap, the greater the heterogeneity and the less transferable the average result.

For example, a meta-analysis yielded the pooled RR of 0.75, with a 95% CI of 0.55–1.02, but the PI spans 0.32–1.73. The null [RR = 1] lies comfortably inside the PI, warning that future trials could plausibly observe either benefit or harm.

It is worth noting that non-statistical audiences [e.g., clinicians] absorb magnitudes better when expressed in absolute risks. Suppose the baseline admission rate is 20%. A pooled RR of 0.75 translates to *four fewer admissions per 100 patients on average*. The PI indicates that a new study might see anything from *14 fewer to 15 more admissions per 100* – an easy straightforward sentence that preserves uncertainty while remaining concrete. Forest-plot graphics can underscore the message by shading the PI as a light-grey [or sometimes red] bar behind the familiar blue diamond of the pooled effect, an approach now common in BMJ and JAMA reviews (IntHout et al., 2016).

3.2. Practical Uses of Prediction Interval

The Grading of Recommendations Assessment, Development and Evaluation [GRADE] approach treats inconsistency across studies as a major threat to certainty (Guyatt et al., 2008). Because a 95% PI reflects both the pooled average and the between-study variance [τ^2], it instantly reveals whether heterogeneity is simply "statistical chatter" or clinically meaningful. If the PI stretches across the line of no effect – indicating that a future, comparable study could plausibly show either important benefit or important harm – the panel should downgrade the certainty of evidence for inconsistency, even when the confidence interval [CI] around the mean is narrowly "significant" (Guyatt et al., 2008). Using the PI avoids overconfidence that can creep in when one focuses on the CI alone.

Besides, the prediction interval is also helpful in planning for a new RCT. Investigators designing the next RCT often struggle with which effect size to plug into a sample-size estimator. In light of this, the 95% PI offers a ready-made three-point menu:

- Optimistic scenario: lower PI limit [greatest benefit]
- Average scenario: pooled point estimate
- Pessimistic scenario: upper PI limit [least benefit or even harm]

This strategy grounds sample-size assumptions in observed evidence rather than hunches. It also flags when powering a study on the mean effect might be futile because a sizeable part of the predicted range crosses the null, prompting either larger sample sizes or a refined research question.

Another practical use of PI is to reconcile "contradictory" findings. Beginners are often puzzled when a meta-analysis reports a statistically significant pooled effect [CI excludes the null] yet the PI still crosses it. Far from signalling a computational error, this scenario illustrates a fundamental truth: the average benefit co-exists with genuine variability across settings, populations, or implementation fidelity (IntHout et al., 2016). Displaying both intervals side-by-side clarifies that what "works on average" may not work everywhere, directing attention toward exploring effect modifiers instead of questioning the analysis pipeline.

4. ADDRESSING HETEROGENEITY

High between-study heterogeneity is more than a statistical footnote; it threatens the very credibility of a pooled estimate. When I^2 soars or the prediction interval stretches from benefit to harm, systematic reviewers should move beyond "report and shrug" and instead take a structured set of follow-up steps – each of which must be transparently disclosed under PRISMA 2020 (Page, McKenzie, Bossuyt et al, 2021; Page, Moher, Bossuyt et al, 2021).

The first response is to look for plausible *sources* of variability, i.e., where does the variability come from? Subgroup analyses or meta-regression can test whether differences in population age, dosage, geographical region or study quality explain the spread of effects. This will be discussed later in this book. When a modifier such as continent or intervention intensity convincingly accounts for the dispersion, reviewers should present separate pooled estimates rather than a single grand mean, because that provides end-users with context-specific expectations. PRISMA 2020 explicitly instructs authors to "describe any methods used to explore possible causes of heterogeneity," making this investigative step a reporting obligation, not a luxury (Page, McKenzie, Bossuyt et al, 2021; Page, Moher, Bossuyt et al, 2021).

If exploratory analyses do not pinpoint a culprit, the next task is to ask whether a single outlying study is distorting the picture. Influence diagnostics – including Baujat plots and Viechtbauer-Cheung influence charts – visualise how much each study contributes to the heterogeneity statistic [Q] and how strongly it pulls the pooled effect. Studies that sit in the upper-right quadrant merit close scrutiny: they may be methodologically atypical, measure a subtly different construct, or simply contain a data error. Conducting a *leave-one-out* sensitivity run reveals whether the conclusions stand without that study. Importantly, potential outliers should not be discarded reflexively; any exclusion must be justified in clinical or methodological terms, as Viechtbauer himself emphasized (Viechtbauer, ; Viechtbauer & Cheung, 2010).

A third line of defence is to examine *model fit*. Comparing fixed-effect and random-effects models can be enlightening: a dramatic swing in the pooled estimate or its uncertainty bands suggests that the data do not conform to a "common-effect" assumption. Even within random-effects analyses, authors should report the τ^2 estimate [and its uncertainty] alongside I^2, because τ^2 quantifies *how much* the true effects differ, whereas I^2 merely states that they differ. Prediction intervals translate that variance back onto the original outcome scale, making the practical impact of heterogeneity explicit (Borenstein, 2023; Borenstein, 2024).

When studies in a meta-analysis show quite different results that researchers really explain through statistics or logical reasoning, it is a good idea to be more cautious about the overall average effect. The GRADE system, for example, specifically points to this kind of unexplained difference as a reason to have less confidence in how consistent the evidence is (Guyatt et al., 2008). Another option for researchers is to decide that combining the data statistically may not be the right approach. In these situations, describing the findings of each study separately, perhaps with a visual display like a forest plot that shows each study's result without an overall average, can be a clearer way to present the evidence. Both of these strategies help to remind those making decisions not to rely too heavily on a single average result when there's a lot of real variation across the individual studies.

Sensitivity analyses provide a final reality check. Excluding studies at high risk of bias, with extreme design features, or based on non-standard outcome definitions can show whether heterogeneity falls and whether the overall interpretation flips. Robust conclusions should not depend on a handful of fragile data points.

Throughout this process, *transparent communication* is paramount. PRISMA 2020 calls for heterogeneity statistics to appear in both the abstract and the main results. By following this protocol of conducting and reporting results, and the pearls that had been discussed in this chapter, reviewers can transform high heterogeneity from a source of confusion into an opportunity for deeper insight and more responsible evidence synthesis (Page, McKenzie, Bossuyt et al, 2021; Page, Moher, Bossuyt et al, 2021).

CONCLUSION AND FUTURE DIRECTIONS

Understanding and addressing heterogeneity remains a cornerstone challenge in meta-analysis. Throughout this chapter, we have explored how metrics like Cochran's Q, I^2, τ^2, H statistic, and prediction intervals can illuminate the complex landscape of between-study variability. These tools, when properly interpreted, transform heterogeneity from a statistical nuisance into valuable information that enriches evidence synthesis.

The journey from detecting heterogeneity to meaningful interpretation requires a systematic approach. First, researchers should recognize that heterogeneity is not merely statistical noise but often reflects genuine clinical or methodological differences between studies. Second, no single metric tells the complete story – Cochran's Q provides a significance test but suffers from power issues, I^2 offers an intuitive percentage but is scale-free, τ^2 quantifies variance in the original metric but lacks context-independent benchmarks, and H statistics distinguish between levels of extreme heterogeneity. Finally, prediction intervals translate statistical heterogeneity into practical implications for future research and clinical decision-making.

Moving forward, several developments in heterogeneity assessment deserve attention. First, there is growing recognition that heterogeneity should be anticipated and planned for, not treated as an unexpected complication. Meta-analysis protocols should pre-specify heterogeneity thresholds that would trigger subgroup analyses or influence the decision to pool results. This shifts heterogeneity from a post-hoc finding to an integral part of design.

Second, the field has been moving beyond simple "high/medium/low" categorizations of I^2 toward more nuanced interpretations that consider the outcome measure, precision of individual studies, and clinical context. Future guidelines should emphasize contextual interpretation rather than rigid cutoffs.

Third, improved visualization techniques are needed. While forest plots effectively display individual studies and the pooled estimate, graphical representations of heterogeneity remain underdeveloped. Future research should explore interactive displays that allow users to visualize how different sources of heterogeneity influence conclusions.

Fourth, Bayesian approaches or machine learning approaches to heterogeneity may offer new ways to identify patterns in heterogeneity that traditional subgroup analyses miss. These methods, though not discussed in this chapter, is worth exploring and could help detect complex interactions between study characteristics that explain variability in effects.

Finally, the field should make effort to further encourage reporting practices. Despite PRISMA 2020 guidelines, heterogeneity measures are often reported inconsistently or incompletely. Prediction interval was recommended to be rountinely reported, but it is still not yet a standard practice. Future research should follow standardized reporting guidelines that ensure comprehensive heterogeneity assessment, including confidence intervals for heterogeneity metrics themselves.

In conclusion, heterogeneity should not be viewed as merely a statistical obstacle but as an opportunity to deepen our understanding of intervention effects across contexts. By embracing the complexity that heterogeneity reveals, researchers can produce more nuanced, honest, and ultimately useful evidence syntheses. The future of meta-analysis lies not in pursuing the elusive goal of homogeneity, but in developing better tools to characterize, understand, and communicate the rich variation that exists in real-world evidence. This evolution will require methodological innovation, improved reporting standards, and a shift in mindset from viewing heterogeneity as a problem to recognizing it as a source of insight.

REFERENCES

Borenstein, M. (2023). How to understand and report heterogeneity in a meta-analysis: The difference between I^2 and prediction intervals. *Integrative Medicine Research, 12*(4), 101014. DOI: 10.1016/j.imr.2023.101014 PMID: 38938910

Borenstein, M. (2024). Avoiding common mistakes in meta-analysis: Understanding the distinct roles of Q, I^2, tau^2, and the prediction interval in reporting heterogeneity. *Research Synthesis Methods, 15*(2), 354–368. DOI: 10.1002/jrsm.1678 PMID: 37940120

Borenstein, M., Hedges, L. V., Higgins, J. P. T., & Rothstein, H. R. (2021). *Introduction to meta-analysis*. John Wiley & Sons. DOI: 10.1002/9781119558378

Dat, T. V., Tu, V. L., Quan, N. K., Minh, N. H., Trung, T. D., Le, T. N., & Huy, N. T. (2023). Telepharmacy: A systematic review of field application, benefits, limitations, and applicability during the COVID-19 pandemic. *Telemedicine Journal and e-Health, 29*(2), 209–221. PMID: 35758766

Guddat, C., Grouven, U., Bender, R., & Skipka, G. (2012). A note on the graphical presentation of prediction intervals in random-effects meta-analyses. *Systematic Reviews, 1*(1), 34. Advance online publication. DOI: 10.1186/2046-4053-1-34 PMID: 22839660

Guyatt, G. H., Oxman, A. D., Vist, G. E., Kunz, R., Falck-Ytter, Y., Alonso-Coello, P., & Schünemann, H. J. (2008). GRADE: An emerging consensus on rating quality of evidence and strength of recommendations. *BMJ (Clinical Research Ed.), 336*(7650), 924–926. DOI: 10.1136/bmj.39489.470347.AD PMID: 18436948

Higgins, J. P. T., Thomas, J., Chandler, J., Cumpston, M., Li, T., Page, M. J., ... (2024). *Cochrane handbook for systematic reviews of interventions* (Version 6.5, updated August 2024). Cochrane. https://training.cochrane.org/handbook

Higgins, J. P. T., & Thompson, S. G. (2002). Quantifying heterogeneity in a meta-analysis. *Statistics in Medicine*, *21*(11), 1539–1558. DOI: 10.1002/sim.1186 PMID: 12111919

IntHout, J., Ioannidis, J. P. A., Rovers, M. M., & Goeman, J. J.IntHout. (2016). Plea for routinely presenting prediction intervals in meta-analysis. *BMJ Open*, *6*(7), e010247. DOI: 10.1136/bmjopen-2015-010247 PMID: 27406637

Ioannidis, J. P. A., Patsopoulos, N. A., & Evangelou, E. (2007). Uncertainty in heterogeneity estimates in meta-analyses. *BMJ (Clinical Research Ed.)*, *335*(7626), 914–916. DOI: 10.1136/bmj.39343.408449.80 PMID: 17974687

Minh, L. H. N., Tawfik, G. M., Ghozy, S., Hashan, M. R., Nam, N. H., Linh, L. K., Abdelrahman, S. A. M., Quynh, T. T. H., Khoi Quan, N., Nhat Le, T., Ibrahim, H. Y., El-Nile, M. O., Kamel, A. M. A., Giang, H. T. N., & Huy, N. T. (2021). Feto-maternal outcomes of breastfeeding during pregnancy: A systematic review and meta-analysis. *Journal ofTropical Pediatrics*, *67*(6), fmab097. DOI: 10.1093/tropej/fmab097 PMID: 34962568

Page, M. J., McKenzie, J. E., Bossuyt, P. M., Boutron, I., Hoffmann, T. C., Mulrow, C. D., Shamseer, L., Tetzlaff, J. M., Akl, E. A., Brennan, S. E., Chou, R., Glanville, J., Grimshaw, J. M., Hróbjartsson, A., Lalu, M. M., Li, T., Loder, E. W., Mayo-Wilson, E., McDonald, S., & Moher, D. (2021). The PRISMA 2020 statement: An updated guideline for reporting systematic reviews. *BMJ (Clinical Research Ed.)*, *372*(71), n71. Advance online publication. DOI: 10.1136/bmj.n71 PMID: 33782057

Page, M. J., Moher, D., Bossuyt, P. M., Boutron, I., Hoffmann, T. C., Mulrow, C. D., Shamseer, L., Tetzlaff, J. M., Akl, E. A., Brennan, S. E., Chou, R., Glanville, J., Grimshaw, J. M., Hróbjartsson, A., Lalu, M. M., Li, T., Loder, E. W., Mayo-Wilson, E., McDonald, S., & McKenzie, J. E. (2021). PRISMA 2020 explanation and elaboration: Updated guidance and exemplars for reporting systematic reviews. *BMJ (Clinical Research Ed.)*, *372*(160), n160. Advance online publication. DOI: 10.1136/bmj.n160 PMID: 33781993

Riley, R. D., Higgins, J. P. T., & Deeks, J. J. (2011). Interpretation of random effects meta-analyses. *BMJ (Clinical Research Ed.)*, *342*(feb10 2), d549. DOI: 10.1136/bmj.d549 PMID: 21310794

Viechtbauer, W., & Cheung, M. W. L. (2010). Outlier and influence diagnostics for meta-analysis. *Research Synthesis Methods*, *1*(2), 112–125. DOI: 10.1002/jrsm.11 PMID: 26061377

Compilation of References

Adeva, J. J. G., Pikatza, J. M., Ubeda, M., & Ansuategi, E. (2014). Automatic text classification to support systematic reviews in medicine. *Expert Systems with Applications*, *41*(4), 1498–1508. DOI: 10.1016/j.eswa.2013.08.047

Al Noman, A., Sarkar, O., Mita, T. M., Siddika, K., & Afrose, F. (2024). Simplifying the concept of level of evidence in lay language for all aspects of learners: A brief review. *Intelligent Pharmacy*, *2*(2), 270–273. DOI: 10.1016/j.ipha.2023.11.002

Altman, D. (2013). Twenty years of meta-analysis and evidence synthesis methods: A personal reflection. *Cochrane Database of Systematic Reviews*, 2–11.

Andermann, A., Blancquaert, I., Beauchamp, S., & Déry, V. (2008). Revisiting Wilson and Jungner in the genomic age: A review of screening criteria over the past 40 years. *Bulletin of the World Health Organization*, *86*(4), 317–319. DOI: 10.2471/BLT.07.050112 PMID: 18438522

Arends, L. R., Hamza, T. H., van Houwelingen, J. C., Heijenbrok-Kal, M. H., Hunink, M. G. M., & Stijnen, T. (2008). Bivariate Random Effects Meta-Analysis of ROC Curves. *Medical Decision Making*, *28*(5), 621–638. DOI: 10.1177/0272989X08319957 PMID: 18591542

Arksey, H., & O'Malley, L. (2005). Scoping studies: Towards a methodological framework. *International Journal of Social Research Methodology*, *8*(1), 19–32. DOI: 10.1080/1364557032000119616

Aromataris, E., Fernandez, R. S., Godfrey, C., Holly, C., Khalil, H., & Tungpunkom, P. (2014). *Methodology for JBI umbrella reviews*. Joanna Briggs Institute.

Aromataris, E., Fernandez, R., Godfrey, C. M., Holly, C., Khalil, H., & Tungpunkom, P. (2015). Summarizing systematic reviews: Methodological development, conduct and reporting of an umbrella review approach. *JBI Evidence Implementation, 13*(3), 132–140. DOI: 10.1097/XEB.0000000000000055 PMID: 26360830

Aromataris, E., & Munn, Z. (Eds.). (2020). *JBI manual for evidence synthesis*. JBI., DOI: 10.46658/JBIMES-24-01

Aslam, S., & Emmanuel, P. (2010). Formulating a researchable question: A critical step for facilitating good clinical research. *Indian Journal of Sexually Transmitted Diseases and AIDS, 31*(1), 47–50. DOI: 10.4103/0253-7184.69003 PMID: 21808439

Ballard, M., & Montgomery, P. (2017). Risk of bias in overviews of reviews. *Research Synthesis Methods, 8*(1), 92–108. DOI: 10.1002/jrsm.1229 PMID: 28074553

Barbour, R. S. (2001). Checklists for improving rigour in qualitative research: A case of the tail wagging the dog? *BMJ (Clinical Research Ed.), 322*(7294), 1115–1117. DOI: 10.1136/bmj.322.7294.1115 PMID: 11337448

Barnett-Page, E., & Thomas, J. (2009). Methods for the synthesis of qualitative research: A critical review. *BMC Medical Research Methodology, 9*(1), 59. Advance online publication. DOI: 10.1186/1471-2288-9-59 PMID: 19671152

Behringer, W., & Freund, Y. (2021). Clinical translation of diagnostic studies: Pitfalls of the usual reported characteristics. *European Journal of Emergency Medicine, 28*(3), 165–166. DOI: 10.1097/MEJ.0000000000000830 PMID: 33899776

Belbasis, L., Bellou, V., & Ioannidis, J. P. A. (2022). Conducting umbrella reviews. *BMJ Medicine*, *1*(1), e000071. DOI: 10.1136/bmjmed-2021-000071 PMID: 36936579

Beller, E., Glasziou, P., Altman, D. G., Hopewell, S., Bastian, H., Chalmers, I., Gøtzsche, P. C., Lasserson, T. J., & Tovey, D. (2013). PRISMA for Abstracts: Reporting Systematic reviews in journal and conference abstracts. *PLoS Medicine*, *10*(4), e1001419. DOI: 10.1371/journal.pmed.1001419 PMID: 23585737

Bentley, J. A., Thoburn, J. W., & Stewart, D. (2019). Cultural adaptations of mental health interventions for refugee populations. *The Journal of Nervous and Mental Disease*, *207*(12), 1044–1052. DOI: 10.1097/NMD.0000000000001092

Bernardo, W. M. (2017). PRISMA statement and PROSPERO. *International Braz J Urol*, *43*(3), 383–384. DOI: 10.1590/s1677-5538.ibju.2017.03.02 PMID: 28520335

Bettany-Saltikov, J. (2010). Learning how to undertake a systematic review: Part 1. *Nursing Standard (Royal College of Nursing (Great Britain): 1987), 24*(50), 47–55; quiz 56. DOI: 10.7748/ns2010.08.24.50.47.c7939

Bettany-Saltikov, J. (2010). Learning how to undertake a systematic review: Part 1. *Nursing Standard*, *24*(50), 47–55. DOI: 10.7748/ns.24.50.47.s49 PMID: 20865948

Bhattacherjee, A. (2012). *Social Science Research: Principles, Methods, and Practices*. University of South Florida.

Bonell, C., Jamal, F., Melendez-Torres, G. J., & Cummins, S. (2015). 'Dark logic': Theorising the harmful consequences of public health interventions. *Journal of Epidemiology and Community Health*, *69*(1), 95–98. DOI: 10.1136/jech-2014-204671 PMID: 25403381

Boote, J. D., Dalgleish, M., Freeman, J., Jones, Z., Miles, M., & Rodgers, H. (2012). But is it a question worth asking? A reflective case study describing how public involvement can lead to researchers' ideas being abandoned. *Health Expectations*. PMID: 22646745Booth, A., Sutton, A., & Papaioannou, D. (2016). *Systematic approaches to a successful literature review*. SAGE Publications.

Booth, A. (2006). Clear and present questions: Formulating questions for evidence based practice. *Library Hi Tech*, *24*(3), 355–368. DOI: 10.1108/07378830610692127

Booth, A., Moore, G., Flemming, K., Garside, R., Rollins, N., Tunçalp, Ö., & Noyes, J. (2019). Taking account of context in systematic reviews and guidelines considering a complexity perspective. *BMJ Global Health*, *4*(Suppl 1), e000840. DOI: 10.1136/bmjgh-2018-000840 PMID: 30775011

Borenstein, M. (2023). How to understand and report heterogeneity in a meta-analysis: The difference between I^2 and prediction intervals. *Integrative Medicine Research*, *12*(4), 101014. DOI: 10.1016/j.imr.2023.101014 PMID: 38938910

Borenstein, M. (2024). Avoiding common mistakes in meta-analysis: Understanding the distinct roles of Q, I^2, tau^2, and the prediction interval in reporting heterogeneity. *Research Synthesis Methods*, *15*(2), 354–368. DOI: 10.1002/jrsm.1678 PMID: 37940120

Borenstein, M., Hedges, L. V., Higgins, J. P. T., & Rothstein, H. R. (2021). *Introduction to meta-analysis*. John Wiley & Sons. DOI: 10.1002/9781119558378

Bowden, V. R., & Purper, C. (2022). Types of reviews—Part 3: Literature review, integrative review, scoping review. *Pediatric Nursing*, *48*(2), 97–101. DOI: 10.62116/PNJ.2022.48.2.97

Brooke, B. S., Schwartz, T. A., & Pawlik, T. M. (2021). MOOSE reporting guidelines for meta-analyses of observational studies. *JAMA Surgery*, *156*(8), 787–788. DOI: 10.1001/jamasurg.2021.0522 PMID: 33825847

Brunner, J., & Austin, P. C. (2009). Inflation of Type I error rate in multiple regression when independent variables are measured with error. *The Canadian Journal of Statistics*, *37*(1), 33–46. https://doi.org/https://doi.org/10.1002/cjs.10004. DOI: 10.1002/cjs.10004

Campbell, J. M. K. M., Ding S, Carmody DP, Hakonsen SJ, Jadotte YT, White S, Munn Z. (2020). *Chapter 9: Diagnostic test accuracy systematic reviews*. JBI. *Cochrane Methods GRADEing*. Retrieved May 9 from https://methods.cochrane.org/gradeing/research

Campbell, F., Tricco, A. C., Munn, Z., Pollock, D., Saran, A., Sutton, A., White, H., & Khalil, H. (2023). Mapping reviews, scoping reviews, and evidence and gap maps (EGMs): The same but different—The "big picture" review family. *Systematic Reviews*, *12*(1), 45. Advance online publication. DOI: 10.1186/s13643-023-02178-5 PMID: 36918977

Campbell, M., Katikireddi, S. V., Sowden, A., & Thomson, H. (2019). Lack of transparency in reporting narrative synthesis of quantitative data: A methodological assessment of systematic reviews. *Journal of Clinical Epidemiology*, *105*, 1–9. DOI: 10.1016/j.jclinepi.2018.08.019 PMID: 30196129

Cant, R., Ryan, C., & Kelly, M. A. (2022). A nine-step pathway to conduct an umbrella review of literature. *Nurse Education in Practice*, *32*(2), 31–34. DOI: 10.1111/nae2.12039

Centre for Reviews and Dissemination. (2009). *Systematic Reviews: CRD's guidance for undertaking reviews in health care*. University of York. https://www.york.ac.uk/inst/crd/guidance.htm

Chalmers, I., Bracken, M. B., Djulbegovic, B., Garattini, S., Grant, J., Gülmezoglu, A. M., Howells, D. W., Ioannidis, J. P. A., & Oliver, S. (2014). How to increase value and reduce waste when research priorities are set. *Lancet*, *383*(9912), 156–165. DOI: 10.1016/S0140-6736(13)62229-1 PMID: 24411644

Chen, C. (2006). CiteSpace II: Detecting and visualizing emerging trends and transient patterns in scientific literature. *Journal of the American Society for Information Science and Technology*, *57*(3), 359–377. DOI: 10.1002/asi.20317

Chen, Y. F., Takwoingi, Y., & Grove, A. (2023). Mapping review as a subtype of scoping review: Differentiating between mapping evidence and presentation of evidence as maps. *Journal of Clinical Epidemiology*, *157*, 110–111. DOI: 10.1016/j.jclinepi.2023.02.017 PMID: 36801486

Choi, G. J., & Kang, H. (2023). Introduction to umbrella reviews as a useful evidence-based practice. *Journal of Lipid and Atherosclerosis*, *12*(1), 3–11. DOI: 10.12997/jla.2023.12.1.3 PMID: 36761061

Choi, S. W., & Lam, D. M. H. (2016). Funnels for publication bias—Have we lost the plot? *Anaesthesia*, *71*(3), 338–341. DOI: 10.1111/anae.13355 PMID: 26693674

Choi, W., Song, S., Ock, S., Kim, C., Lee, J. B., Chang, W., & Kim, S. (2014). Duplicate publication of articles used in meta-analysis in Korea. *SpringerPlus*, *3*(1), 182. DOI: 10.1186/2193-1801-3-182 PMID: 24808996

Clark, E., Burkett, K., & Stanko-Lopp, D. (2009). Let evidence guide every new decision (LEGEND): An evidence evaluation system for point-of-care clinicians and guideline development teams. *Journal of Evaluation in Clinical Practice*, *15*(6), 1054–1060. DOI: 10.1111/j.1365-2753.2009.01314.x PMID: 20367705

Clarke, A. M., Kuosmanen, T., & Barry, M. M. (2015). A systematic review of online youth mental health promotion and prevention interventions. *Journal of Youth and Adolescence*, *44*(1), 90–113. DOI: 10.1007/s10964-014-0165-0 PMID: 25115460

Cohen, J. F., Korevaar, D. A., Altman, D. G., Bruns, D. E., Gatsonis, C. A., Hooft, L., Irwig, L., Levine, D., Reitsma, J. B., de Vet, H. C. W., & Bossuyt, P. M. (2016). STARD 2015 guidelines for reporting diagnostic accuracy studies: Explanation and elaboration. *BMJ Open*, *6*(11), e012799. DOI: 10.1136/bmjopen-2016-012799 PMID: 28137831

Collaboration for Environmental Evidence. (2013). *Guidelines for Systematic Review and Evidence Synthesis in Environmental Management.* Version 4.2. Environmental Evidence: www.environmentalevidence.org/Documents/Guidelines/Guidelines4.2.pdf

Collins, G. S., Reitsma, J. B., Altman, D. G., & Moons, K. G. M. (2015). Transparent reporting of a multivariable prediction model for individual prognosis or diagnosis (TRIPOD): The TRIPOD Statement. *BMC Medicine, 13*(1), 1. DOI: 10.1186/s12916-014-0241-z PMID: 25563062

Conroy, G. (2023). How ChatGPT and other AI tools could disrupt scientific publishing. *Nature, 622*(7982), 234–236. DOI: 10.1038/d41586-023-03144-w PMID: 37817033

Cooke, A., Smith, D., & Booth, A. (2012). Beyond PICO: The SPIDER tool for qualitative evidence synthesis. *Qualitative Health Research, 22*(10), 1435–1443. DOI: 10.1177/1049732312452938 PMID: 22829486

Cooper, H., & Koenka, A. C. (2012). The overview of reviews: Unique challenges and opportunities when research syntheses are the principal elements of new integrative scholarship. *The American Psychologist, 67*(6), 446–462. DOI: 10.1037/a0027119 PMID: 22352742

Cooper, I. D. (2016). What is a "mapping study"? *Journal of the Medical Library Association: JMLA, 104*(1), 76–78. DOI: 10.3163/1536-5050.104.1.013 PMID: 26807058

Covidence—Better systematic review management. (n.d.). Retrieved April 5, 2024, from https://www.covidence.org/

Crawford, C., Brown, L. M., Costello, R. B., & Deuster, P. A. (2022). Select dietary supplement ingredients for preserving and protecting the immune system in healthy individuals: A systematic review. *Nutrients, 14*(21), 4604. DOI: 10.3390/nu14214604 PMID: 36364865

Creswell, J. W., & Creswell, J. D. (2017). *Research Design: Qualitative, Quantitative, and Mixed Methods Approaches.* SAGE Publications.

Critical Appraisal Skills Programme. (n.d.). *CASP checklists*. https://casp-uk.net/casp-tools-checklists/

Critical Appraisal tools. (n.d.). [Web Page]. Retrieved April 5, 2024, from https://www.cebm.ox.ac.uk/resources/ebm-tools/critical-appraisal-tools

Cumpston, M. S., Brennan, S. E., Ryan, R., & McKenzie, J. E. (2023). Synthesis methods other than meta-analysis were commonly used but seldom specified: Survey of systematic reviews. *Journal of Clinical Epidemiology*, *156*, 42–52. DOI: 10.1016/j.jclinepi.2023.02.003 PMID: 36758885

Dat, T. V., Tu, V. L., Quan, N. K., Minh, N. H., Trung, T. D., & Le, T. N.. (2023). Telepharmacy: A systematic review of field application, benefits, limitations, and applicability during the COVID-19 pandemic. *Telemedicine Journal and e-Health*, *29*(2), 209–221. DOI: 10.1089/tmj.2022.0043 PMID: 35758766

De Caro, W., Marucci, A. R., Lancia, L., & Sansoni, J. (2016). Umbrella reviews. In *Evidence synthesis with overviews of reviews and meta-epidemiologic studies* (pp. 150–180). Springer.

De Cassai, A., Boscolo, A., Zarantonello, F., Pettenuzzo, T., Sella, N., Geraldini, F., Munari, M., & Navalesi, P. (2023). Enhancing study quality assessment: An in-depth review of risk of bias tools for meta-analysis. *Journal of Anesthesia, Analgesia and Critical Care*, *3*(1), 44. Advance online publication. DOI: 10.1186/s44158-023-00129-z PMID: 37932825

Deeks, J. J., Bossuyt, P. M., Leeflang, M. M., & Takwoingi, Y. (2023). *Cochrane handbook for systematic reviews of diagnostic test accuracy*. John Wiley & Sons. DOI: 10.1002/9781119756194

DeMars, M. M., & Perruso, C. (2022). MeSH and text-word search strategies: Precision, recall, and their implications for library instruction. *Journal of the Medical Library Association: JMLA*, *110*(1), 23–33. DOI: 10.5195/jmla.2022.1283 PMID: 35210959

Department for Environment Food and Rural Affairs. (2015). *Emerging tools and techniques to deliver timely and cost effective evidence reviews*. DEFRA.

DerSimonian, R., & Laird, N. (1986). Meta-analysis in clinical trials. *Controlled Clinical Trials, 7*(3), 177–188. DOI: 10.1016/0197-2456(86)90046-2 PMID: 3802833

Dias, S., Sutton, A. J., Welton, N. J., & Ades, A. E. (2013). Evidence Synthesis for Decision Making 3: Heterogeneity—Subgroups, Meta-Regression, Bias, and Bias-Adjustment. *Medical Decision Making, 33*(5), 618–640. DOI: 10.1177/0272989X13485157 PMID: 23804507

Dixon-Woods, M., Agarwal, S., Jones, D., Young, B., & Sutton, A. (2005). Synthesising qualitative and quantitative evidence: A review of possible methods. *Journal of Health Services Research & Policy, 10*(1), 45–53. DOI: 10.1177/135581960501000110 PMID: 15667704

Dol, J., Campbell-Yeo, M., Tomblin Murphy, G., Aston, M., McMillan, D., Gahagan, J., & Richardson, B. (2019). Parent-targeted postnatal educational interventions in low- and middle-income countries: A scoping review and critical analysis. *International Journal of Nursing Studies, 94*, 60–73. DOI: 10.1016/j.ijnurstu.2019.03.011 PMID: 30933873

Doyle, L. H. (2003). Synthesis through meta-ethnography: Paradoxes, enhancements, and possibilities. *Qualitative Research, 3*(3), 321–344. DOI: 10.1177/1468794103033003

Drucker, A. M., Fleming, P., & Chan, A.-W. (2016). Research techniques made simple: Assessing risk of bias in systematic reviews. *The Journal of Investigative Dermatology, 136*(11), e109–e114. DOI: 10.1016/j.jid.2016.08.021 PMID: 27772550

Dukic, V., & Gatsonis, C. (2003). Meta-analysis of diagnostic test accuracy assessment studies with varying number of thresholds. *Biometrics, 59*(4), 936–946. DOI: 10.1111/j.0006-341X.2003.00108.x PMID: 14969472

Eccleston, C., Fisher, E., & Vowles, K. E.. (2017). Psychological therapies (Internet-delivered) for the management of chronic pain in adults. *Cochrane Database of Systematic Reviews*, 2. PMID: 24574082

Edwards, P., Clarke, M., DiGuiseppi, C., Pratap, S., Roberts, I., & Wentz, R. (2002). Identification of randomized controlled trials in systematic reviews: Accuracy and reliability of screening records. *Statistics in Medicine*, *21*(11), 1635–1640. DOI: 10.1002/sim.1190 PMID: 12111924

Eloundou, T., Manning, S., Mishkin, P., & Rock, D. (2023). Gpts are gpts: An early look at the labor market impact potential of large language models. arXiv preprint arXiv:2303.10130.

Eriksen, M. B., & Frandsen, T. F. (2018). The impact of patient, intervention, comparison, outcome (PICO) as a search strategy tool on literature search quality: A systematic review. *Journal of the Medical Library Association: JMLA*, *106*(4), 420–431. DOI: 10.5195/jmla.2018.345 PMID: 30271283

Erwin, E. J., Brotherson, M. J., & Summers, J. A. (2011). Understanding qualitative metasynthesis. *Journal of Early Intervention*, *33*(3), 186–200. DOI: 10.1177/1053815111425493

Ewald, H., Klerings, I., Wagner, G., Heise, T. L., Stratil, J. M., Lhachimi, S. K., Hemkens, L. G., Gartlehner, G., Armijo-Olivo, S., & Nussbaumer-Streit, B. (2022). Searching two or more databases decreased the risk of missing relevant studies: A metaresearch study. *Journal of Clinical Epidemiology*, *149*, 154–164. https://doi.org/https://doi.org/10.1016/j.jclinepi.2022.05.022. DOI: 10.1016/j.jclinepi.2022.05.022 PMID: 35654269

Eysenck, H. J. (1994). Systematic Reviews: Meta-analysis and its problems. *BMJ (Clinical Research Ed.)*, *309*(6957), 789–792. DOI: 10.1136/bmj.309.6957.789 PMID: 7950571

Flemming, K., & Noyes, J. (2021). Qualitative evidence synthesis: Where are we at? *International Journal of Qualitative Methods*, *20*, 1609406921993276. DOI: 10.1177/1609406921993276

Frampton, G. K., Livoreil, B., & Petrokofsky, G. (2017). Eligibility screening in evidence synthesis of environmental management topics. *Environmental Evidence*, *6*(1), 27. DOI: 10.1186/s13750-017-0102-2

France, E. F., Uny, I., Ring, N., Turley, R. L., Maxwell, M., Duncan, E. A. S., Jepson, R. G., Roberts, R. J., & Noyes, J. (2019). A methodological systematic review of meta-ethnography conduct to articulate the complex analytical phases. *BMC Medical Research Methodology*, *19*(1), 35. DOI: 10.1186/s12874-019-0670-7 PMID: 30777031

Fryback, D. G., & Thornbury, J. R. (1991). The efficacy of diagnostic imaging. *Medical Decision Making*, *11*(2), 88–94. DOI: 10.1177/0272989X9101100203 PMID: 1907710

Fu, H., Wang, L., Zhang, W., Lu, J., & Yang, M. (2023). Diagnostic test accuracy of ultrasound for sarcopenia diagnosis: A systematic review and meta-analysis. *Journal of Cachexia, Sarcopenia and Muscle*, *14*(1), 57–70. DOI: 10.1002/jcsm.13149 PMID: 36513380

Fusar-Poli, P., & Radua, J. (2018). Ten simple rules for conducting umbrella reviews. *Evidence-Based Mental Health*, *21*(3), 95–100. DOI: 10.1136/ebmental-2018-300014 PMID: 30006442

Gagnon, M.-P., Desmartis, M., & Labrecque, M.. (2016). Implementation of telehealth in an interprofessional rehabilitation university clinic: A mixed-method study. *Telemedicine Journal and e-Health*, *22*(11), 951–959. DOI: 10.1089/tmj.2015.0221

Galvin, R., Cusack, T., & Stokes, E. (2011). To what extent are family members and friends involved in physiotherapy and the delivery of exercises to people with stroke? A qualitative study. *Disability and Rehabilitation*, *33*(15–16), 1395–1404. DOI: 10.3109/09638288.2010.533244

Gartlehner, G., Affengruber, L., Titscher, V., Noel-Storr, A., Dooley, G., Ballarini, N., & König, F. (2020). Single-reviewer abstract screening missed 13 percent of relevant studies. *Journal of Clinical Epidemiology*, *121*, 20–28. DOI: 10.1016/j.jclinepi.2020.01.005 PMID: 31972274

Gates, M., Gates, A., Guitard, S., Pollock, M., & Hartling, L. (2020). Guidance for overviews of reviews continues to accumulate. *Systematic Reviews*, *9*(1), 254. Advance online publication. DOI: 10.1186/s13643-020-01509-0 PMID: 33148319

Geneen, L. J., Moore, R. A., & Clarke, C.. (2017). Physical activity and exercise for chronic pain in adults: An overview of Cochrane reviews. *Cochrane Database of Systematic Reviews*, 4.

Gillman, L. N., & Wright, S. D. (2010). Mega mistakes in meta-analyses: Devil in the detail. *Ecology*, *91*(9), 2550–2552. DOI: 10.1890/09-0339.1 PMID: 20957947

Glanville, J., Kotas, E., Featherstone, R., & Dooley, G. (2020). Which are the most sensitive search filters to identify randomized controlled trials in MEDLINE? *Journal of the Medical Library Association: JMLA*, *108*(4), 556–563. DOI: 10.5195/jmla.2020.912 PMID: 33013212

Glas, A. S., Lijmer, J. G., Prins, M. H., Bonsel, G. J., & Bossuyt, P. M. (2003). The diagnostic odds ratio: A single indicator of test performance. *Journal of Clinical Epidemiology*, *56*(11), 1129–1135. DOI: 10.1016/S0895-4356(03)00177-X PMID: 14615004

Godino, L. (2023). How to structure Microsoft Excel documents for systematic reviews. *Nurse Researcher*, *31*(1), 40–46. DOI: 10.7748/nr.2023.e1866 PMID: 36856031

González-Suárez, C., Worley, A., Grimmer-Somers, K., & Dones, V. (2013). School-based interventions on childhood obesity: A meta-analysis. *American Journal of Preventive Medicine*, *44*(5), 517–523. DOI: 10.1016/j.amepre.2012.12.007 PMID: 19840696

Gopalakrishna, G., Mustafa, R. A., Davenport, C., Scholten, R. J., Hyde, C., Brozek, J., Schünemann, H. J., Bossuyt, P. M. M., Leeflang, M. M. G., & Langendam, M. W. (2014). Applying Grading of Recommendations Assessment, Development and Evaluation (GRADE) to diagnostic tests was challenging but doable. *Journal of Clinical Epidemiology*, *67*(7), 760–768. DOI: 10.1016/j.jclinepi.2014.01.006 PMID: 24725643

Gough, D. (2013). Meta-narrative and realist reviews: Guidance, rules, publication standards and quality appraisal. *BMC Medicine*, *11*(1), 22. Advance online publication. DOI: 10.1186/1741-7015-11-22 PMID: 23360691

Gough, D., Oliver, S., & Thomas, J. (2017). *An Introduction to Systematic Reviews*. SAGE Publications.

GRADE Handbook. (n.d.). Retrieved April 5, 2024, from https://training.cochrane.org/resource/grade-handbook

GRADEpro GDT. https://www.gradepro.org/

Grant, M. J., & Booth, A. (2009). A typology of reviews: An analysis of 14 review types and associated methodologies. *Health Information and Libraries Journal*, *26*(2), 91–108. DOI: 10.1111/j.1471-1842.2009.00848.x PMID: 19490148

Greco, T., Zangrillo, A., Biondi-Zoccai, G., & Landoni, G. (2013). Meta-analysis: Pitfalls and hints. *Heart, Lung and Vessels*, *5*(4), 219–225. PMID: 24364016

Greenhalgh, T. (2014). *How to Read a Paper: The Basics of Evidence-Based Medicine*. John Wiley & Sons.

Green, S., & Higgins, J. P. T. (2011). *Cochrane Handbook for Systematic Reviews of Interventions*. John Wiley & Sons.

Guddat, C., Grouven, U., Bender, R., & Skipka, G. (2012). A note on the graphical presentation of prediction intervals in random-effects meta-analyses. *Systematic Reviews*, *1*(1), 34. Advance online publication. DOI: 10.1186/2046-4053-1-34 PMID: 22839660

Gutierrez-Arias, R., Pieper, D., Lunny, C., Torres-Castro, R., Aguilera-Eguía, R., & Seron, P. (2023). Strategies used to manage overlap of primary study data by exercise-related overviews: Protocol for a systematic methodological review. *BMJ Open*, *13*(4), e069906. DOI: 10.1136/bmjopen-2022-069906 PMID: 37080626

Guyatt, G. H., Oxman, A. D., Vist, G. E., Kunz, R., Falck-Ytter, Y., Alonso-Coello, P., & Schünemann, H. J. (2008). GRADE: An emerging consensus on rating quality of evidence and strength of recommendations. *BMJ (Clinical Research Ed.), 336*(7650), 924–926. DOI: 10.1136/bmj.39489.470347.AD PMID: 18436948

Hannes, K., & Macaitis, K. (2012). A move to more systematic and transparent approaches in qualitative evidence synthesis: Update on a review of published papers. *Qualitative Research, 12*(4), 402–422. DOI: 10.1177/1468794111432992

Harrison, H., Griffin, S. J., Kuhn, I., & Usher-Smith, J. A. (2020). Software tools to support title and abstract screening for systematic reviews in healthcare: An evaluation. *BMC Medical Research Methodology, 20*(1), 7. DOI: 10.1186/s12874-020-0897-3 PMID: 31931747

Hart, C. (1998). *Doing a Literature Review: Releasing the Social Science Research Imagination*. SAGE Publications.

Hartling, L., Featherstone, R., Nuspl, M., Shave, K., Dryden, D. M., & Vandermeer, B. (2016). The contribution of databases to the results of systematic reviews: A cross-sectional study. *BMC Medical Research Methodology, 16*(1), 127. Advance online publication. DOI: 10.1186/s12874-016-0232-1 PMID: 27670136

Hennessy, E. A., & Johnson, B. T. (2020). Examining overlap of included studies in meta-reviews. *Research Synthesis Methods, 11*(1), 134–145. DOI: 10.1002/jrsm.1390 PMID: 31823513

Higgins, J. P. T., Thomas, J., Chandler, J., Cumpston, M., Li, T., Page, M. J., ... (2024). *Cochrane handbook for systematic reviews of interventions* (Version 6.5, updated August 2024). Cochrane. https://training.cochrane.org/handbook

Higgins, J. P. T., Altman, D. G., Gøtzsche, P. C., Juni, P., Moher, D., Oxman, A. D., Savovic, J., Schulz, K. F., Weeks, L., & Sterne, J. A. C. (2011). The Cochrane Collaboration's tool for assessing risk of bias in randomised trials. *BMJ (Clinical Research Ed.), 343*(oct18 2), d5928. DOI: 10.1136/bmj.d5928 PMID: 22008217

Higgins, J. P. T., Lasserson, T., Chandler, J., Tovey, D., & Churchill, R. (2016). *Methodological expectations of Cochrane intervention reviews*. Cochrane.

Higgins, J. P. T., & Thompson, S. G. (2002). Quantifying heterogeneity in a meta-analysis. *Statistics in Medicine*, *21*(11), 1539–1558. DOI: 10.1002/sim.1186 PMID: 12111919

Hong, Q. N., & Pluye, P. (2018). A conceptual framework for critical appraisal in systematic mixed studies reviews. *Journal of Mixed Methods Research*, *13*(4), 1558689818770058. Advance online publication. DOI: 10.1177/1558689818770058

Horwitz, R. I. (1995). "Large-scale randomized evidence: Large, simple trials and overviews of trials": Discussion. A clinician's perspective on meta-analyses. *Journal of Clinical Epidemiology*, *48*(1), 41–44. DOI: 10.1016/0895-4356(94)00171-L PMID: 7853046

Hosseini, M.-S., Jahanshahlou, F., Akbarzadeh, M. A., Zarei, M., & Vaez-Gharamaleki, Y. (2024). Formulating research questions for evidence-based studies. *Journal of Medicine, Surgery, and Public Health*, *2*, 100046. https://doi.org/https://doi.org/10.1016/j.glmedi.2023.100046. DOI: 10.1016/j.glmedi.2023.100046

Hulley, S. B., Cummings, S. R., & Browner, W. S. (Eds.). (2007). *Designing Clinical Research: An Epidemiological Approach* (4th ed., pp. 14–22). Lippincott Williams & Wilkins.

Hulley, S. B., Cummings, S. R., Browner, W. S., Grady, D. G., & Newman, T. B. (2013). *Designing Clinical Research*. Lippincott Williams & Wilkins.

Iared, W., Mourão, J. E., Puchnick, A., Soma, F., & Shigueoka, D. C. (2022). Angioplasty versus stenting for subclavian artery stenosis. *The Cochrane Library*, *2022*(2), CD008461. Advance online publication. DOI: 10.1002/14651858.CD008461.pub4 PMID: 35187653

Institute of Medicine (US) Committee on Standards for Systematic Reviews of Comparative Effectiveness Research. (2011). *Finding What Works in Health Care: Standards for Systematic Reviews* (Eden, J., Levit, L., Berg, A., & Morton, S., Eds.). National Academies Press., DOI: 10.17226/13059

IntHout, J., Ioannidis, J. P. A., Rovers, M. M., & Goeman, J. J.IntHout. (2016). Plea for routinely presenting prediction intervals in meta-analysis. *BMJ Open*, *6*(7), e010247. DOI: 10.1136/bmjopen-2015-010247 PMID: 27406637

Ioannidis, J. P. A. (2009). Integration of evidence from multiple meta-analyses: A primer on umbrella reviews, treatment networks and multiple treatments meta-analyses. *Canadian Medical Association Journal*, *181*(8), 488–493. DOI: 10.1503/cmaj.081086 PMID: 19654195

Ioannidis, J. P. A. (2016). The mass production of redundant, misleading, and conflicted systematic reviews and meta-analyses. *The Milbank Quarterly*, *94*(3), 485–514. DOI: 10.1111/1468-0009.12210 PMID: 27620683

Ioannidis, J. P. A. (2017). Next-generation systematic reviews: Prospective meta-analysis, individual-level data, networks and umbrella reviews. *British Journal of Sports Medicine*, *51*(20), 1456–1458. DOI: 10.1136/bjsports-2017-097621 PMID: 28223307

Ioannidis, J. P. A., Patsopoulos, N. A., & Evangelou, E. (2007). Uncertainty in heterogeneity estimates in meta-analyses. *BMJ (Clinical Research Ed.)*, *335*(7626), 914–916. DOI: 10.1136/bmj.39343.408449.80 PMID: 17974687

Jabali, S. H., Yazdani, S., Pourasghari, H., & Maleki, M. (2024). From bench to policy: A critical analysis of models for evidence-informed policymaking in healthcare. *Frontiers in Public Health*, *12*, 1264315. DOI: 10.3389/fpubh.2024.1264315 PMID: 38596514

James, K. L., Randall, N. P., & Haddaway, N. R. (2016). A methodology for systematic mapping in environmental sciences. *Environmental Evidence*, *5*(1), 7. Advance online publication. DOI: 10.1186/s13750-016-0059-6

Kausik, B. (2023). Long tails & the impact of GPT on labor. Available at *SSRN* 4525008. DOI: 10.2139/ssrn.4525008

Kavalieratos, D., Corbelli, J., Zhang, D., Dionne-Odom, J. N., Ernecoff, N. C., Hanmer, J., Hoydich, Z. P., Ikejiani, D. Z., Klein-Fedyshin, M., Zimmermann, C., Morton, S. C., Arnold, R. M., Heller, L., & Schenker, Y. (2016). Association between palliative care and patient and caregiver outcomes: A systematic review and meta-analysis. *Journal of the American Medical Association, 316*(20), 2104–2114. DOI: 10.1001/jama.2016.16840 PMID: 27893131

Khalil, H., Campbell, F., Danial, K., Pollock, D., Munn, Z., Welsh, V., Saran, A., Hoppe, D., & Tricco, A. C. (2024). Advancing the methodology of mapping reviews: A scoping review. *Research Synthesis Methods, 15*(3), 384–397. DOI: 10.1002/jrsm.1694 PMID: 38169156

Khalil, H., & Tricco, A. C. (2022). Differentiating between mapping reviews and scoping reviews in the evidence synthesis ecosystem. *Journal of Clinical Epidemiology, 149*, 175–182. DOI: 10.1016/j.jclinepi.2022.05.012 PMID: 35636593

Khoi Quan, N., & Liamputtong, P. (2023). Systematic review and evidence synthesis in public health. In Liamputtong, P. (Ed.), *Handbook of social sciences and global public health* (pp. 1–25). Springer., DOI: 10.1007/978-3-030-96778-9_143-1

Klerings, I., Weinhandl, A. S., & Thaler, K. J. (2015). Information overload in healthcare: Too much of a good thing? *Zeitschrift für Evidenz, Fortbildung und Qualität im Gesundheitswesen, 109*(4–5), 285–290. DOI: 10.1016/j.zefq.2015.06.005 PMID: 26354128

Kneale, D., Thomas, J., & Harris, K. (2015). Developing and optimising the use of logic models in systematic reviews: Exploring practice and good practice in the use of programme theory in reviews. *PLoS One, 10*(11), e0142187. DOI: 10.1371/journal.pone.0142187 PMID: 26575182

Langan, D. (2022). Assessing HeterogeneityHeterogeneity in Random-Effects Meta-analysis. In Evangelou, E., & Veroniki, A. A. (Eds.), *Meta-Research: Methods and Protocols* (pp. 67–89). Springer US., DOI: 10.1007/978-1-0716-1566-9_4

Langford, R., Bonell, C. P., Jones, H. E., Pouliou, T., Murphy, S. M., Waters, E., Komro, K. A., Gibbs, L. F., Magnus, D., & Campbell, R. (2014). The WHO Health Promoting School framework for improving the health and well-being of students and their academic achievement. *Cochrane Database of Systematic Reviews, 2014*(4), 4. DOI: 10.1002/14651858.CD008958.pub2 PMID: 24737131

Langhorne, P., Bernhardt, J., & Kwakkel, G. (2011). Stroke rehabilitation. *Lancet, 377*(9778), 1693–1702. DOI: 10.1016/S0140-6736(11)60325-5 PMID: 21571152

Laver, K., Lange, B., & George, S.. (2020). Virtual reality for stroke rehabilitation. *Cochrane Database of Systematic Reviews*, 11.

Learn more about INPLASY. Retrieved May 9 from https://inplasy.com/learn-more-about-inplasy/

Leeflang, M. M. (2014). Systematic reviews and meta-analyses of diagnostic test accuracy. *Clinical Microbiology and Infection, 20*(2), 105–113. DOI: 10.1111/1469-0691.12474 PMID: 24274632

Lee, J., Kim, K. W., Choi, S. H., Huh, J., & Park, S. H. (2015). Systematic Review and Meta-Analysis of Studies Evaluating Diagnostic Test Accuracy: A Practical Review for Clinical Researchers-Part II. Statistical Methods of Meta-Analysis. *Korean Journal of Radiology, 16*(6), 1188–1196. DOI: 10.3348/kjr.2015.16.6.1188 PMID: 26576107

Lefebvre, C., Glanville, J., Briscoe, S., Featherstone, R., Littlewood, A., Metzendorf, M.-I., Noel-Storr, A., Paynter, R., Rader, T., Thomas, J., & Wieland, L. S. (2023). *Chapter 4: Searching for and selecting studies*. In: Higgins JPT, Thomas J, Chandler J, Cumpston M, Li T, Page MJ, Welch VA (editors). Cochrane Handbook for Systematic Reviews of Interventions version 6.4 (updated October 2023). Cochrane. https://training.cochrane.org/handbook

Levac, D., Colquhoun, H., & O'Brien, K. K. (2010). Scoping studies: Advancing the methodology. *Implementation Science : IS*, *5*(1), 69. Advance online publication. DOI: 10.1186/1748-5908-5-69 PMID: 20854677

Lewin, S., Bohren, M., Rashidian, A., Munthe-Kaas, H., Glenton, C., Colvin, C. J., Garside, R., Noyes, J., Booth, A., Tunçalp, Ö., Wainwright, M., Flottorp, S., Tucker, J. D., & Carlsen, B. (2018). Applying GRADE-CERQual to qualitative evidence synthesis findings—paper 2: How to make an overall CERQual assessment of confidence and create a Summary of Qualitative Findings table. *Implementation Science : IS*, *13*(1), 10. DOI: 10.1186/s13012-017-0689-2 PMID: 29384082

Liberati, A., Altman, D. G., Tetzlaff, J., Mulrow, C. D., Gøtzsche, P. C., Ioannidis, J. P. A., Clarke, M., Devereaux, P., Kleijnen, J., & Moher, D. (2009). The PRISMA statement for reporting systematic reviews and meta-analyses of studies that evaluate healthcare interventions: Explanation and elaboration. *BMJ (Clinical Research Ed.)*, *339*(1), b2700. DOI: 10.1136/bmj.b2700 PMID: 19622552

Lisy, K., & Porritt, K. (2016). Narrative synthesis. *International Journal of Evidence-Based Healthcare*, *14*(4), 201. DOI: 10.1097/01.XEB.0000511348.97198.8c

Liu, Z., Al Amer, F. M., Xiao, M., Xu, C., Furuya-Kanamori, L., Hong, H., Siegel, L., & Lin, L. (2023). The normality assumption on between-study random effects was questionable in a considerable number of Cochrane meta-analyses. *BMC Medicine*, *21*(1), 112. DOI: 10.1186/s12916-023-02823-9 PMID: 36978059

Liu, Z., Yao, Z., Li, C., Liu, X., Chen, H., & Gao, C. (2013). A step-by-step guide to the systematic review and meta-analysis of diagnostic and prognostic test accuracy evaluations. *British Journal of Cancer*, *108*(11), 2299–2303. DOI: 10.1038/bjc.2013.185 PMID: 23695015

Livoreil, B., Glanville, J., Haddaway, N., Bayliss, H., Bethel, A., De Lachapelle, F. F., Robalino, S., Savilaakso, S., Zhou, W., Petrokofsky, G., & Frampton, G. K. (2017). Systematic searching for environmental evidence using multiple tools and sources. *Environmental Evidence*, *6*(1), 23. DOI: 10.1186/s13750-017-0099-6

Loades, M. E., Chatburn, E., Higson-Sweeney, N., Reynolds, S., Shafran, R., Brigden, A., Linney, C., McManus, M. N., Borwick, C., & Crawley, E. (2020). Rapid systematic review: The impact of social isolation and loneliness on the mental health of children and adolescents in the context of COVID-19. *Journal of the American Academy of Child and Adolescent Psychiatry*, *59*(11), 1218–1239.e3. DOI: 10.1016/j.jaac.2020.05.009 PMID: 32504808

López-López, J. A., Rubio-Aparicio, M., & Sánchez-Meca, J. (2022). Overviews of reviews: Concept and development. *Psicothema*, *34*(2), 175–181. DOI: 10.7334/psicothema2021.586 PMID: 35485529

Lorenc, T., Petticrew, M., Welch, V., & Tugwell, P. (2013). What types of interventions generate inequalities? Evidence from systematic reviews. *Journal of Epidemiology and Community Health*, *67*(2), 190–193. DOI: 10.1136/jech-2012-201257 PMID: 22875078

Lunny, C., Brennan, S. E., McDonald, S., & McKenzie, J. E. (2017). Toward a comprehensive evidence map of overview methods. *Systematic Reviews*, *6*, 231. Advance online publication. DOI: 10.1186/s13643-017-0617-1 PMID: 29162130

Lunny, C., Pieper, D., Thabet, P., & Kanji, S. (2021). Managing overlap of primary study results across systematic reviews. *BMC Medical Research Methodology*, *21*(1), 140. Advance online publication. DOI: 10.1186/s12874-021-01269-y PMID: 34233615

Macaskill, P. (2004). Empirical Bayes estimates generated in a hierarchical summary ROC analysis agreed closely with those of a full Bayesian analysis. *Journal of Clinical Epidemiology*, *57*(9), 925–932. DOI: 10.1016/j.jclinepi.2003.12.019 PMID: 15504635

Maggio, L. A., Larsen, K., Thomas, A., Costello, J. A., & Artino, A. R.Jr. (2021). Scoping reviews in medical education: A scoping review. *Medical Education, 55*(6), 689–700. DOI: 10.1111/medu.14431 PMID: 33300124

Magwood, O., Riddle, A., Petkovic, J., Lytvyn, L., Khabsa, J., Atwere, P., Akl, E. A., Campbell, P., Welch, V., Smith, M., Mustafa, R. A., Limburg, H., Dans, L. F., Skoetz, N., Grant, S., Concannon, T. W., & Tugwell, P. (2022). PROTOCOL: Barriers and facilitators to stakeholder engagement in health guideline development: A qualitative evidence synthesis. *Campbell Systematic Reviews, 18*(2), e1237. DOI: 10.1002/cl2.1237 PMID: 36911345

Major, C. H., & Savin-Baden, M. (2010). *An introduction to qualitative research synthesis*. Routledge., DOI: 10.4324/9780203497555

Mathew, M. J. (2024). Literature search in systematic reviews: How much is good enough? *Clinical Epidemiology and Global Health, 25*, 101485. https://doi.org/https://doi.org/10.1016/j.cegh.2023.101485. DOI: 10.1016/j.cegh.2023.101485

Mayer, S. J. (2018). *The Dynamic Student Development Meta-Theory: A New Model for Student Success*. Routledge.

McInnes, M. D. F., Moher, D., Thombs, B. D., McGrath, T. A., Bossuyt, P. M., Clifford, T., Cohen, J. F., Deeks, J. J., Gatsonis, C., Hooft, L., Hunt, H. A., Hyde, C. J., Korevaar, D. A., Leeflang, M. M. G., Macaskill, P., Reitsma, J. B., Rodin, R., Rutjes, A. W. S., Salameh, J.-P., & Willis, B. H. (2018). Preferred Reporting Items for a Systematic Review and Meta-analysis of Diagnostic Test Accuracy Studies: The PRISMA-DTA Statement. *Journal of the American Medical Association, 319*(4), 388–396. DOI: 10.1001/jama.2017.19163 PMID: 29362800

Mckenzie, J., Brennan, S., Ryan, R., Thomson, H., Johnston, R., & Thomas, J. (2019). Chapter 3: Defining the criteria for including studies and how they will be grouped for the synthesis. In *Cochrane Handbook for Systematic Reviews of Interventions*. DOI: 10.1002/9781119536604.ch3

McLuckie, A., Landers, A. L., Curran, J. A., Cann, R., Carrese, D. H., Nolan, A., Corrigan, K., & Carrey, N. J. (2019). A scoping review of mental health prevention and intervention initiatives for infants and preschoolers at risk for socio-emotional difficulties. *Systematic Reviews*, *8*(1), 183. Advance online publication. DOI: 10.1186/s13643-019-1043-3 PMID: 31337424

Meline, T. (2006). Selecting Studies for Systemic Review: Inclusion and Exclusion Criteria. *Contemporary Issues in Communication Science and Disorders*, *33*(Spring), 21–27. DOI: 10.1044/cicsd_33_S_21

Merrick, H., Driver, H., Main, C., Kenny, R. P. W., Richmond, C., Allard, A., Bola, K., Morris, C., Parr, J. R., Pearson, F., & Pennington, L. (2023). Impacts of health care service changes implemented due to COVID-19 on children and young people with long-term disability: A mapping review. *Developmental Medicine and Child Neurology*, *65*(7), 885–899. DOI: 10.1111/dmcn.15503 PMID: 36649197

Miake-Lye, I. M., Hempel, S., Shanman, R., & Shekelle, P. G. (2016). What is an evidence map? *Systematic Reviews*, *5*(1), 28. Advance online publication. DOI: 10.1186/s13643-016-0204-x PMID: 26864942

Miles, M. B., Huberman, A. M., & Saldaña, J. (2014). *Qualitative Data Analysis: A Methods Sourcebook*. SAGE Publications.

Minh, L. H. N., Tawfik, G. M., Ghozy, S., Hashan, M. R., Nam, N. H., Linh, L. K., & Nhat Le, T. (2021). Feto-maternal outcomes of breastfeeding during pregnancy: A systematic review and meta-analysis. *Journal of Tropical Pediatrics*, *67*(6), fmab097. DOI: 10.1093/tropej/fmab097 PMID: 34962568

Munn, Z., Moola, S., Lisy, K., Riitano, D., & Tufanaru, C. (2020). Methodological guidance for systematic reviews of observational epidemiological studies reporting prevalence and cumulative incidence data. *International Journal of Evidence-Based Healthcare*, *18*(3), 147–153. DOI: 10.1097/XEB.0000000000000054 PMID: 26317388

Munn, Z., Peters, M. D. J., Stern, C., Tufanaru, C., McArthur, A., & Aromataris, E. (2018). Systematic review or scoping review? *BMC Medical Research Methodology*, *18*(1), 143. DOI: 10.1186/s12874-018-0611-x PMID: 30453902

Munn, Z., Pollock, D., Khalil, H., Alexander, L., Mclnerney, P., Godfrey, C. M., Peters, M., & Tricco, A. C. (2022). What are scoping reviews? *JBI Evidence Synthesis*, *20*(4), 950–952. DOI: 10.11124/JBIES-21-00483 PMID: 35249995

Murad, M. H., Asi, N., Alsawas, M., & Alahdab, F. (2016). New evidence pyramid. *Evidence-Based Medicine*, *21*(4), 125–127. DOI: 10.1136/ebmed-2016-110401 PMID: 27339128

Nasser, M., Ueffing, E., Welch, V., & Tugwell, P. (2013). An equity lens can ensure an equity-oriented approach to agenda setting and priority setting of Cochrane Reviews. *Journal of Clinical Epidemiology*, *66*(5), 511–521. DOI: 10.1016/j.jclinepi.2012.11.013 PMID: 23477991

National Academies of Sciences. E., and Medicine; Health and Medicine Division; Board on the Health of Select Populations; Board on Health Care Services; Committee on the Evidence Base for Genetic Testing. (2017). *An Evidence Framework for Genetic Testing*. National Academies Press (US). https://doi.org/DOI: 10.17226/24632

Neumann, I., Brignardello-Petersen, R., Wiercioch, W., Carrasco-Labra, A., Cuello, C., Akl, E., Mustafa, R. A., Al-Hazzani, W., Etxeandia-Ikobaltzeta, I., Rojas, M. X., Falavigna, M., Santesso, N., Brozek, J., Iorio, A., Alonso-Coello, P., & Schünemann, H. J. (2015). The GRADE evidence-to-decision framework: A report of its testing and application in 15 international guideline panels. *Implementation Science : IS*, *11*(1), 93. DOI: 10.1186/s13012-016-0462-y PMID: 27417219

Neuman, W. L. (2014). *Social Research Methods: Qualitative and Quantitative Approaches*. Pearson Education.

Noyes, J., Lewin, S., & Rees, R. (2011). *A guide to synthesising qualitative research for researchers undertaking health technology assessments and systematic reviews*. University of Stirling. https://www.stir.ac.uk/research/hub/publication/817639

Oliver, S., Armes, D. G., & Gyte, G. (2009). Public involvement influences a national research agenda. *Patient, 2*(3), 179–190. DOI: 10.2165/11314860-000000000-00000 PMID: 22273169

Olsen, M., Zhelev, Z., Hunt, H., Peters, J. L., Bossuyt, P., & Hyde, C. (2019). Use of test accuracy study design labels in NICE's diagnostic guidance. *Diagnostic and Prognostic Research, 3*(1), 17. DOI: 10.1186/s41512-019-0062-9 PMID: 31517065

Ouzzani, M., Hammady, H., Fedorowicz, Z., & Elmagarmid, A. (2016). Rayyan—A web and mobile app for systematic reviews. *Systematic Reviews, 5*(1), 210. Advance online publication. DOI: 10.1186/s13643-016-0384-4 PMID: 27919275

Pae, C. U. (2015). Why systematic review rather than narrative review? *Psychiatry Investigation, 12*(3), 417. DOI: 10.4306/pi.2015.12.3.417 PMID: 26207140

Page, M. J., McKenzie, J. E., Bossuyt, P. M., Boutron, I., Hoffmann, T. C., Mulrow, C. D., Shamseer, L., Tetzlaff, J. M., Akl, E. A., Brennan, S. E., Chou, R., Glanville, J., Grimshaw, J. M., Hróbjartsson, A., Lalu, M. M., Li, T., Loder, E. W., Mayo-Wilson, E., McDonald, S., & Moher, D. (2021). The PRISMA 2020 statement: An updated guideline for reporting systematic reviews. *BMJ (Clinical Research Ed.), 372*(71), n71. Advance online publication. DOI: 10.1136/bmj.n71 PMID: 33782057

Page, M. J., Moher, D., Bossuyt, P. M., Boutron, I., Hoffmann, T. C., Mulrow, C. D., Shamseer, L., Tetzlaff, J. M., Akl, E. A., Brennan, S. E., Chou, R., Glanville, J., Grimshaw, J. M., Hróbjartsson, A., Lalu, M. M., Li, T., Loder, E. W., Mayo-Wilson, E., McDonald, S., & McKenzie, J. E. (2021). PRISMA 2020 explanation and elaboration: Updated guidance and exemplars for reporting systematic reviews. *BMJ (Clinical Research Ed.)*, *372*(160), n160. Advance online publication. DOI: 10.1136/bmj.n160 PMID: 33781993

Palese, A., Mansutti, I., Visintini, E., Caruzzo, D., Moreale, R., Longhini, J., & Danielis, M. (2022). Framing the time while designing and conducting reviews: A focused mapping review and synthesis. *Journal of Clinical Nursing*, *31*(23–24), 3523–3534. DOI: 10.1111/jocn.16180 PMID: 34936157

Papatheodorou, S. I., & Evangelou, E. (2022). Umbrella reviews: What they are and why we need them. In *Methods in molecular biology* (Vol. 2345, pp. 135–146). Springer., DOI: 10.1007/978-1-0716-1566-9_8

Patel, A., Cooper, N., Freeman, S., & Sutton, A. (2021). Graphical enhancements to summary receiver operating characteristic plots to facilitate the analysis and reporting of meta-analysis of diagnostic test accuracy data. *Research Synthesis Methods*, *12*(1), 34–44. DOI: 10.1002/jrsm.1439 PMID: 32706182

Patino, C. M., & Ferreira, J. C. (2018). Inclusion and exclusion criteria in research studies: Definitions and why they matter. *Jornal Brasileiro de Pneumologia : Publicaçao Oficial da Sociedade Brasileira de Pneumologia e Tisilogia*, *44*(2), 84. DOI: 10.1590/s1806-37562018000000088 PMID: 29791550

Patton, M. Q. (2015). *Qualitative Research & Evaluation Methods: Integrating Theory and Practice*. SAGE Publications.

Pearson, K. (2010). Report on certain enteric fever inoculation statistics (Original work published 1904). *BMJ*. Republished in *The James Lind Library*. https://www.jameslindlibrary.org/pearson-k-1904/

Persson, O., Danell, R., & Schneider, J. W. (2009). How to use Bibexcel for various types of bibliometric analysis. Celebrating scholarly communication studies: A Festschrift for Olle Persson at his 60th Birthday, 5(2009), 9-24.

Petersen, K., & Ali, N. B. (2011). Identifying Strategies for Study Selection in Systematic Reviews and Maps. In 2011 International Symposium on Empirical Software Engineering and Measurement. 2011 5th International Symposium on Empirical Software Engineering and Measurement (ESEM). IEEE. DOI: 10.1109/ESEM.2011.46

Peters, M. D. J., Godfrey, C. M., Khalil, H., McInerney, P., Parker, D., & Soares, C. B. (2015). Guidance for conducting systematic scoping reviews. *International Journal of Evidence-Based Healthcare*, *13*(3), 141–146. DOI: 10.1097/XEB.0000000000000050 PMID: 26134548

Peters, M. D. J., Marnie, C., Colquhoun, H., Garritty, C. M., Hempel, S., Horsley, T., Langlois, E. V., Lillie, E., O'Brien, K. K., Tunçalp, ., Wilson, M. G., Zarin, W., & Tricco, A. C. (2021). Scoping reviews: Reinforcing and advancing the methodology and application. *Systematic Reviews*, *10*(1), 263. Advance online publication. DOI: 10.1186/s13643-021-01821-3 PMID: 34625095

Peterson, J., Pearce, P. F., Ferguson, L. A., & Langford, C. A. (2017). Understanding scoping reviews. *Journal of the American Association of Nurse Practitioners*, *29*(1), 12–16. DOI: 10.1002/2327-6924.12380 PMID: 27245885

Petticrew, M., & Roberts, H. (2008). *Systematic Reviews in the Social Sciences: A Practical Guide*. John Wiley & Sons.

Pham, M. T., Rajić, A., Greig, J. D., Sargeant, J. M., Papadopoulos, A., & McEwen, S. A. (2014). A scoping review of scoping reviews. *Research Synthesis Methods*, *5*(4), 371–385. DOI: 10.1002/jrsm.1123 PMID: 26052958

Pieper, D., Buechter, R., Jerinic, P., & Eikermann, M. (2012). Overviews of reviews often have limited rigor. *Journal of Clinical Epidemiology*, *65*(12), 1267–1273. DOI: 10.1016/j.jclinepi.2012.06.015 PMID: 22959594

Pieper, D., & Rombey, T. (2022). Where to prospectively register a systematic review. *Systematic Reviews*, *11*(1), 8. DOI: 10.1186/s13643-021-01877-1 PMID: 34998432

Pigott, T. D., & Polanin, J. R. (2020). Methodological Guidance Paper: High-Quality Meta-Analysis in a Systematic Review. *Review of Educational Research*, *90*(1), 24–46. DOI: 10.3102/0034654319877153

Polanin, J. R., Pigott, T., Espelage, D. L., & Grotpeter, J. K. (2019). Best practice guidelines for abstract screening large-evidence systematic reviews and meta-analyses. *Research Synthesis Methods*, *10*(3), 330–342. DOI: 10.1002/jrsm.1354

Pollock, M., Fernandes, R. M., Becker, L. A., Pieper, D., & Hartling, L. (2020). Overviews of reviews. In *Cochrane handbook for systematic reviews of interventions* (Version 6). Cochrane.

Pollock, D., Peters, M. D. J., Khalil, H., McInerney, P., Alexander, L., Tricco, A. C., Evans, C., de Moraes, É. B., Godfrey, C. M., Pieper, D., Saran, A., Stern, C., & Munn, Z. (2023). Recommendations for the extraction, analysis, and presentation of results in scoping reviews. *JBI Evidence Synthesis*, *21*(3), 520–532. DOI: 10.11124/JBIES-22-00123 PMID: 36081365

Pollock, M., Fernandes, R. M., Becker, L. A., Featherstone, R., & Hartling, L. (2016). What guidance is available for researchers conducting overviews of reviews of healthcare interventions? *Systematic Reviews*, *5*(1), 190. Advance online publication. DOI: 10.1186/s13643-016-0367-5 PMID: 27842604

Pollock, M., Fernandes, R. M., Newton, A. S., Scott, S. D., & Hartling, L. (2019). A decision tool to help researchers make decisions about including systematic reviews in overviews of reviews of healthcare interventions. *Systematic Reviews*, *8*(1), 29. Advance online publication. DOI: 10.1186/s13643-018-0768-8 PMID: 30670086

Popay, J., Roberts, H., & Sowden, A.. (2006). *Guidance on the conduct of narrative synthesis in systematic reviews*. ESRC Methods Programme.

PRISMA for Diagnostic Test Accuracy. Retrieved May 9 from https://www.prisma-statement.org/dta

PRISMA Team. (n.d.). *PRISMA flow diagram generator (Shiny app)*. https://estech.shinyapps.io/prisma_flowdiagram/

PROSPERO. (n.d.). *International prospective register of systematic reviews*. Centre for Reviews and Dissemination, University of York. https://www.crd.york.ac.uk/prospero/

Purgato, M., Tol, W. A., & Bass, J. K.. (2018). Psychosocial interventions for children exposed to traumatic events in low- and middle-income countries: A systematic review and meta-analysis. *Journal of Child Psychology and Psychiatry, and Allied Disciplines*, *59*(7), 767–783. DOI: 10.1111/jcpp.12851

Rayyan – Intelligent Systematic Review—Rayyan. (n.d.). Retrieved April 5, 2024, from https://www.rayyan.ai/

Rethlefsen, M. L., & Page, M. J. (2022). PRISMA 2020 and PRISMA-S. *Journal of the Medical Library Association: JMLA*, *110*(2), 253–257. DOI: 10.5195/jmla.2022.1449 PMID: 35440907

Richardson, W. S., Wilson, M. C., Nishikawa, J., & Hayward, R. S. (1995). The well-built clinical question: A key to evidence-based decisions. *ACP Journal Club*, *123*(3), A12–A13. DOI: 10.7326/ACPJC-1995-123-3-A12 PMID: 7582737

Riley, R. D., Higgins, J. P. T., & Deeks, J. J. (2011). Interpretation of random effects meta-analyses. *BMJ (Clinical Research Ed.)*, *342*(feb10 2), d549. DOI: 10.1136/bmj.d549 PMID: 21310794

Rodgers, M., Sowden, A., Petticrew, M., Arai, L., Roberts, H., Britten, N., & Popay, J. (2009). Testing methodological guidance on the conduct of narrative synthesis in systematic reviews. *Evaluation*, *15*(1), 49–73. DOI: 10.1177/1356389008097871

Rodríguez, L. G., Hogarth, N. J., Zhou, W., Xie, C., Zhang, K., & Putzel, L. (2016). China's conversion of cropland to forest program: A systematic review of the environmental and socioeconomic effects. *Environmental Evidence*, *5*(1), 21. DOI: 10.1186/s13750-016-0071-x

Rutjes, A. W., Reitsma, J. B., Vandenbroucke, J. P., Glas, A. S., & Bossuyt, P. M. (2005). Case-control and two-gate designs in diagnostic accuracy studies. *Clinical Chemistry*, *51*(8), 1335–1341. DOI: 10.1373/clinchem.2005.048595 PMID: 15961549

Salameh, J.-P., Bossuyt, P. M., McGrath, T. A., Thombs, B. D., Hyde, C. J., Macaskill, P., Deeks, J. J., Leeflang, M., Korevaar, D. A., Whiting, P., Takwoingi, Y., Reitsma, J. B., Cohen, J. F., Frank, R. A., Hunt, H. A., Hooft, L., Rutjes, A. W. S., Willis, B. H., Gatsonis, C., & McInnes, M. D. F. (2020). Preferred reporting items for systematic review and meta-analysis of diagnostic test accuracy studies (PRISMA-DTA): Explanation, elaboration, and checklist. *BMJ (Clinical Research Ed.)*, *370*, m2632. DOI: 10.1136/bmj.m2632 PMID: 32816740

Salvador-Oliván, J. A., Marco-Cuenca, G., & Arquero-Avilés, R. (2021). Development of an efficient search filter to retrieve systematic reviews from PubMed. *Journal of the Medical Library Association: JMLA*, *109*(4), 561–570. DOI: 10.5195/jmla.2021.1223 PMID: 34858085

Sanders, T., Noetel, M., Parker, P., Del Pozo Cruz, B., Biddle, S., Ronto, R., Hulteen, R., Parker, R., Thomas, G., De Cocker, K., Salmon, J., Hesketh, K., Weeks, N., Arnott, H., Devine, E., Vasconcellos, R., Pagano, R., Sherson, J., Conigrave, J., & Lonsdale, C. (2023). An umbrella review of the benefits and risks associated with youths' interactions with electronic screens. *Nature Human Behaviour*, *8*(1), 82–99. Advance online publication. DOI: 10.1038/s41562-023-01712-8 PMID: 37957284

Saran, A., & White, H. (2018). Evidence and gap maps: A comparison of different approaches. *Campbell Systematic Reviews*, *14*(1), 1.

Sattar, R., Lawton, R., Panagioti, M., & Johnson, J. (2021). Meta-ethnography in healthcare research: A guide to using a meta-ethnographic approach for literature synthesis. *BMC Health Services Research, 21*(1), 50. Advance online publication. DOI: 10.1186/s12913-020-06049-w PMID: 33419430

Scholten, R. J. P. M., Clarke, M., & Hetherington, J. (2005). The Cochrane Collaboration. *European Journal of Clinical Nutrition, 59*(S1, Suppl 1), S147–S149. DOI: 10.1038/sj.ejcn.1602188 PMID: 16052183

Schroll, J. B., Moustgaard, R., & Gøtzsche, P. C. (2011). Dealing with substantial heterogeneity in Cochrane reviews: Cross-sectional study. *BMC Medical Research Methodology, 11*(1), 22. Advance online publication. DOI: 10.1186/1471-2288-11-22 PMID: 21349195

Shamseer, L., Moher, D., Clarke, M., Ghersi, D., Liberati, A., Petticrew, M., Shekelle, P., Stewart, L. A., & the PRISMA-P Group. (2015). Preferred reporting items for systematic review and meta-analysis protocols (PRISMA-P) 2015: Elaboration and explanation. *BMJ, 349*(jan02 1), g7647–g7647. DOI: 10.1136/bmj.g7647

Shaw, E. K., Howard, J., & West, D. R.. (2018). The role of the practice facilitator in implementing the patient-centered medical home (PCMH). *Journal of the American Board of Family Medicine : JABFM, 31*(5), 762–773. DOI: 10.3122/jabfm.2018.05.170294

Shea, B. J., Grimshaw, J. M., Wells, G. A., Boers, M., Andersson, N., Hamel, C., Porter, A. C., Tugwell, P., Moher, D., & Bouter, L. M. (2007). Development of AMSTAR: A measurement tool to assess the methodological quality of systematic reviews. *BMC Medical Research Methodology, 7*(1), 10. Advance online publication. DOI: 10.1186/1471-2288-7-10 PMID: 17302989

Shea, B. J., Reeves, B. C., Wells, G., Thuku, M., Hamel, C., Moran, J., Moher, D., Tugwell, P., Welch, V., Kristjansson, E., & Henry, D. A. (2017). AMSTAR 2: A critical appraisal tool for systematic reviews. *BMJ (Clinical Research Ed.), 358*, j4008. DOI: 10.1136/bmj.j4008 PMID: 28935701

Shim, S. R. (2022). Meta-analysis of diagnostic test accuracy studies with multiple thresholds for data integration. *Epidemiology and Health*, *44*, e2022083. DOI: 10.4178/epih.e2022083 PMID: 36228672

Simmonds, M. C., & Higgins, J. P. T. (2007). Covariate heterogeneity in meta-analysis: Criteria for deciding between meta-regression and individual patient data. *Statistics in Medicine*, *26*(15), 2982–2999. https://doi.org/https://doi.org/10.1002/sim.2768. DOI: 10.1002/sim.2768 PMID: 17195960

Siontis, K. C., Hernandez-Boussard, T., & Ioannidis, J. P. A. (2013). Overlapping meta-analyses on the same topic. *BMJ (Clinical Research Ed.)*, *347*(jul19 1), f4501. DOI: 10.1136/bmj.f4501 PMID: 23873947

Smith, J. A. (2015). *Qualitative Psychology: A Practical Guide to Research Methods*. SAGE Publications.

Smith, S., Brick, A., O'Hara, S., & Normand, C. (2014). Evidence on the cost and cost-effectiveness of palliative care: A literature review. *Palliative Medicine*, *28*(2), 130–150. DOI: 10.1177/0269216313493466 PMID: 23838378

Smith, V., Devane, D., Begley, C. M., & Clarke, M. (2011). Methodology in conducting a systematic review of systematic reviews of healthcare interventions. *BMC Medical Research Methodology*, *11*(1), 15. Advance online publication. DOI: 10.1186/1471-2288-11-15 PMID: 21291558

Snell, K. I. E., Levis, B., Damen, J. A. A., Dhiman, P., Debray, T. P. A., Hooft, L., Reitsma, J. B., Moons, K. G. M., Collins, G. S., & Riley, R. D. (2023). Transparent reporting of multivariable prediction models for individual prognosis or diagnosis: Checklist for systematic reviews and meta-analyses (TRIPOD-SRMA). *BMJ (Clinical Research Ed.)*, *381*, e073538. DOI: 10.1136/bmj-2022-073538 PMID: 37137496

Software for meta-analysis of DTA studies. https://methods.cochrane.org/sdt/software-meta-analysis-dta-studies

Sonntag, D., Aldinger, L., Bechert, S., Alvarez, M., Groenewolt, A., Krieg, O. D., & Menges, A. (2019). Lightweight segmented timber shell for the Bundesgartenschau in Heilbronn. Paper presented at the Proceedings of IASS Annual Symposia.

Spencer, E. A., & Brassey, J. (2017). Ascertainment bias. In *Catalogue of bias*. https://catalogofbias.org/biases/ascertainment-bias/

Steinhauser, S., Schumacher, M., & Rücker, G. (2016). Modelling multiple thresholds in meta-analysis of diagnostic test accuracy studies. *BMC Medical Research Methodology, 16*(1), 97. DOI: 10.1186/s12874-016-0196-1 PMID: 27520527

Sterne, J. A. C., & Harbord, R. M. (2004). Funnel plots in meta-analysis. *The Stata Journal, 4*(2), 127–141. DOI: 10.1177/1536867X0400400204

Sterne, J. A. C., Savović, J., Page, M. J., Elbers, R. G., Blencowe, N. S., Boutron, I., Cates, C. J., Cheng, H.-Y., Corbett, M. S., Eldridge, S. M., Emberson, J. R., Hernán, M. A., Hopewell, S., Hróbjartsson, A., Junqueira, D. R., Jüni, P., Kirkham, J. J., Lasserson, T., Li, T., & Higgins, J. P. T. (2019). RoB 2: A revised tool for assessing risk of bias in randomised trials. *BMJ (Clinical Research Ed.), 366*, l4898. DOI: 10.1136/bmj.l4898 PMID: 31462531

Sterne, J. A. C., Sutton, A. J., Ioannidis, J. P. A., Terrin, N., Jones, D. R., Lau, J., Carpenter, J., Rucker, G., Harbord, R. M., Schmid, C. H., Tetzlaff, J., Deeks, J. J., Peters, J., Macaskill, P., Schwarzer, G., Duval, S., Altman, D. G., Moher, D., & Higgins, J. P. T. (2011). Recommendations for examining and interpreting funnel plot asymmetry in meta-analyses of randomised controlled trials. *BMJ (Clinical Research Ed.), 343*(jul22 1), d4002. DOI: 10.1136/bmj.d4002 PMID: 21784880

Straus, S., & Moher, D. (2010). Registering systematic reviews. *Canadian Medical Association Journal, 182*(1), 13–14. DOI: 10.1503/cmaj.081849 PMID: 19620270

Students 4 Best Evidence. (n.d.). *Home*. https://s4be.cochrane.org/

Sukhera, J. (2022). Narrative reviews: Flexible, rigorous, and practical. *Journal of Graduate Medical Education*, *14*(4), 414–417. DOI: 10.4300/JGME-D-22-00480.1 PMID: 35991099

Takwoingi, Y., Guo, B., Riley, R. D., & Deeks, J. J. (2017). Performance of methods for meta-analysis of diagnostic test accuracy with few studies or sparse data. *Statistical Methods in Medical Research*, *26*(4), 1896–1911. DOI: 10.1177/0962280215592269 PMID: 26116616

Takwoingi, Y., & Quinn, T. J. (2018). Review of Diagnostic Test Accuracy (DTA) studies in older people. *Age and Ageing*, *47*(3), 349–355. DOI: 10.1093/ageing/afy023 PMID: 29528366

Takwoingi, Y., Riley, R. D., & Deeks, J. J. (2015). Meta-analysis of diagnostic accuracy studies in mental health. *Evidence-Based Mental Health*, *18*(4), 103–109. DOI: 10.1136/eb-2015-102228 PMID: 26446042

Tawfik, G. M., Dila, K. S., Mohamed, M. Y. F., Tam, D. N. H., Kien, N. D., Ahmed, A. M., & Huy, N. T. (2019). A step by step guide for conducting a systematic review and meta-analysis with simulation data. *Tropical Medicine and Health*, *47*(1), 46. DOI: 10.1186/s41182-019-0165-6 PMID: 31388330

Teixeira da Silva, J. A. (2020). The ICMJE recommendations: Challenges in fortifying publishing integrity. *Irish Journal of Medical Science*, *189*(4), 1179–1181. DOI: 10.1007/s11845-020-02227-1 PMID: 32297306

Temel, J. S., Greer, J. A., Muzikansky, A., Gallagher, E. R., Admane, S., Jackson, V. A., Dahlin, C. M., Blinderman, C. D., Jacobsen, J., Pirl, W. F., Billings, J. A., & Lynch, T. J. (2010). Early palliative care for patients with metastatic non-small-cell lung cancer. *The New England Journal of Medicine*, *363*(8), 733–742. DOI: 10.1056/NEJMoa1000678 PMID: 20818875

Teo, L., Van Elswyk, M. E., Lau, C. S., & Shanahan, C. J. (2023). Title-plus-abstract versus title-only screening. *Systematic Reviews*, *12*, 211. Advance online publication. DOI: 10.1186/s13643-023-02374-3 PMID: 37957691

The Open Science Framework. Retrieved May 9 from https://osf.io/

Thomas, B. H., Ciliska, D., Dobbins, M., & Micucci, S. (2004). A process for systematically reviewing the literature: Providing the research evidence for public health nursing interventions. *Worldviews on Evidence-Based Nursing*, *1*(3), 176–184. DOI: 10.1111/j.1524-475X.2004.04006.x PMID: 17163895

Thomas, J., & Harden, A. (2008). Methods for the thematic synthesis of qualitative research in systematic reviews. *BMC Medical Research Methodology*, *8*(1), 45. DOI: 10.1186/1471-2288-8-45 PMID: 18616818

Thompson, S. G., & Higgins, J. P. T. (2002). How should meta-regression analyses be undertaken and interpreted? *Statistics in Medicine*, *21*(11), 1559–1573. DOI: 10.1002/sim.1187 PMID: 12111920

Tol, W. A., Barbui, C., Galappatti, A., Silove, D., Betancourt, T. S., Souza, R., Golaz, A., & van Ommeren, M. (2011). Mental health and psychosocial support in humanitarian settings: Linking practice and research. *Lancet*, *378*(9802), 2126–2132. DOI: 10.1016/S0140-6736(11)61094-5 PMID: 22008428

Tramèr, M. R., Reynolds, D. J., Moore, R. A., & McQuay, H. (1997). Impact of covert duplicate publication on meta-analysis: A case study. *BMJ (Clinical Research Ed.)*, *315*(7109), 635–640. DOI: 10.1136/bmj.315.7109.635 PMID: 9310564

Tricco, A. C., Lillie, E., Zarin, W., O'Brien, K. K., Colquhoun, H., Levac, D., Moher, D., Peters, M. D. J., Horsley, T., Weeks, L., Hempel, S., Akl, E. A., Chang, C., McGowan, J., Stewart, L., Hartling, L., Aldcroft, A., Wilson, M. G., Garritty, C., & Straus, S. E. (2018). PRISMA extension for scoping reviews (PRISMA-ScR). *Annals of Internal Medicine*, *169*(7), 467–473. DOI: 10.7326/M18-0850 PMID: 30178033

Tricco, A. C., Lillie, E., Zarin, W., O'Brien, K., Colquhoun, H., Kastner, M., Levac, D., Ng, C., Sharpe, J. P., Wilson, K., Kenny, M., Warren, R., Wilson, C., Stelfox, H. T., & Straus, S. E. (2016). A scoping review on the conduct and reporting of scoping reviews. *BMC Medical Research Methodology*, *16*(1), 15. Advance online publication. DOI: 10.1186/s12874-016-0116-4 PMID: 26857112

Trikalinos, T. A., Balion, C. M., Coleman, C. I., Griffith, L., Santaguida, P. L., Vandermeer, B., & Fu, R. (2012). Chapter 8: meta-analysis of test performance when there is a "gold standard". *J Gen Intern Med, 27 Suppl 1*(Suppl 1), S56-66. https://doi.org/DOI: 10.1007/s11606-012-2029-1

van der Braak, K., Heus, P., Orelio, C., Netterström-Wedin, F., Robinson, K. A., Lund, H., & Hooft, L. (2023). Perspectives on systematic review protocol registration: A survey amongst stakeholders in the clinical research publication process. *Systematic Reviews*, *12*(1), 234. DOI: 10.1186/s13643-023-02405-z PMID: 38098085

van Houwelingen, H. C., Arends, L. R., & Stijnen, T. (2002). Advanced methods in meta-analysis: Multivariate approach and meta-regression. *Statistics in Medicine*, *21*(4), 589–624. DOI: 10.1002/sim.1040 PMID: 11836738

Verdejo, C., Tapia-Benavente, L., Schuller-Martínez, B., Vergara-Merino, L., Vargas-Peirano, M., & Silva-Dreyer, A. M. (2021). What you need to know about scoping reviews. *Medwave*, *21*(2), e8144. DOI: 10.5867/medwave.2021.02.8144 PMID: 33914717

Veroniki, A. A., Tsokani, S., Agarwal, R., Pagkalidou, E., Rücker, G., Mavridis, D., & Takwoingi, Y. (2022). Diagnostic test accuracy network meta-analysis methods: A scoping review and empirical assessment. *Journal of Clinical Epidemiology*, *146*, 86–96. https://doi.org/https://doi.org/10.1016/j.jclinepi.2022.02.001. DOI: 10.1016/j.jclinepi.2022.02.001 PMID: 35181490

Viechtbauer, W., & Cheung, M. W. L. (2010). Outlier and influence diagnostics for meta-analysis. *Research Synthesis Methods*, *1*(2), 112–125. DOI: 10.1002/jrsm.11 PMID: 26061377 von Elm, E., Poglia, G., Walder, B., & Tramèr, M. R. (2004). Different patterns of duplicate publication. *Journal of the American Medical Association, 291*(8), 974. DOI: 10.1001/jama.291.8.974 PMID: 14982913

Waffenschmidt, S., Knelangen, M., Sieben, W., Bühn, S., & Pieper, D. (2019). Single screening versus conventional double screening for study selection in systematic reviews. *BMC Medical Research Methodology, 19*(1), 132. Advance online publication. DOI: 10.1186/s12874-019-0782-0 PMID: 31253092

Waffenschmidt, S., Sieben, W., Jakubeit, T., Knelangen, M., Overesch, I., Bühn, S., Pieper, D., Skoetz, N., & Hausner, E. (2023). Increasing the efficiency of study selection for systematic reviews using prioritization tools and a single-screening approach. *Systematic Reviews, 12*(1), 161. DOI: 10.1186/s13643-023-02334-x PMID: 37705060

Wager, E., & Williams, P. (2011). Why and how do journals retract articles? An analysis of Medline retractions 1988-2008. *Journal of Medical Ethics, 37*(9), 567–570. DOI: 10.1136/jme.2010.040964 PMID: 21486985

Wang, J., & Leeflang, M. (2019). Recommended software/packages for meta-analysis of diagnostic accuracy. *Journal of Laboratory and Precision Medicine, 4*, 4. DOI: 10.21037/jlpm.2019.06.01

Waters, E., de Silva-Sanigorski, A., & Hall, B. J.. (2011). Interventions for preventing obesity in children. *Cochrane Database of Systematic Reviews*, 12. PMID: 22161367

Wei, T.-T., Hu, Z.-D., Qin, B.-D., Ma, N., Tang, Q.-Q., Wang, L.-L., Zhou, L., & Zhong, R.-Q. (2016). Diagnostic Accuracy of Procalcitonin in Bacterial Meningitis Versus Nonbacterial Meningitis: A Systematic Review and Meta-Analysis. *Medicine, 95*(11), e3079. https://journals.lww.com/md-journal/fulltext/2016/03150/diagnostic_accuracy_of_procalcitonin_in_bacterial.47.aspx. DOI: 10.1097/MD.0000000000003079 PMID: 26986140

Wells, G. A., Shea, B., O'Connell, D., Peterson, J., Welch, V., Losos, M., & Tugwell, P. (2000). The Newcastle-Ottawa Scale (NOS) for assessing the quality of nonrandomised studies in meta-analyses.

Wells, G. A., Shea, B., O'Connell, D., Peterson, J., Welch, V., Losos, M., ... (n.d.). *The Newcastle–Ottawa Scale (NOS) for assessing the quality of nonrandomised studies in meta-analyses*. https://www.ohri.ca/programs/clinical_epidemiology/oxford.asp

Werner-Seidler, A., Perry, Y., Calear, A. L., Newby, J. M., & Christensen, H. (2017). School-based depression and anxiety prevention programs for young people: A systematic review and meta-analysis. *Clinical Psychology Review*, *51*, 30–47. DOI: 10.1016/j.cpr.2016.10.005 PMID: 27821267

Whiting, P. F., Rutjes, A. W., Westwood, M. E., Mallett, S., Deeks, J. J., Reitsma, J. B., Leeflang, M. M. G., Sterne, J. A. C., & Bossuyt, P. M. (2011). QUADAS-2: A revised tool for the quality assessment of diagnostic accuracy studies. *Annals of Internal Medicine*, *155*(8), 529–536. DOI: 10.7326/0003-4819-155-8-201110180-00009 PMID: 22007046

Whiting, P., Savović, J., Higgins, J. P., Caldwell, D. M., Reeves, B. C., Shea, B., Davies, P., Kleijnen, J., & Churchill, R. (2016). ROBIS: A new tool to assess risk of bias in systematic reviews was developed. *Journal of Clinical Epidemiology*, *69*, 225–234. DOI: 10.1016/j.jclinepi.2015.06.005 PMID: 26092286

Willis, L. D. (2023). Formulating the Research Question and Framing the Hypothesis. *Respiratory Care*, *68*(8), 1180–1185. DOI: 10.4187/respcare.10975 PMID: 37041024

Wong, G., Greenhalgh, T., Westhorp, G., Buckingham, J., & Pawson, R. (2013). RAMESES publication standards: Meta-narrative reviews. *BMC Medicine*, *11*(1), 20. Advance online publication. DOI: 10.1186/1741-7015-11-20 PMID: 23360661

Wu, T., He, S., Liu, J., Sun, S., Liu, K., Han, Q.-L., & Tang, Y. (2023). A brief overview of ChatGPT: The history, status quo and potential future development. IEEE/CAA Journal of Automatica Sinica, 10(5), 1122-1136.

Xu, Z., Jain, S., & Kankanhalli, M. (2024). Hallucination is inevitable: An innate limitation of large language models. arXiv preprint arXiv:2401.11817.

Yimgang, D., Danhoundo, G., Kusi-Appiah, E., Sunder, V., Campbell, S., & Yaya, S. (2021). A scoping review of researchers' involvement in health policy dialogue in Africa. *Systematic Reviews*, *10*(1), 190. Advance online publication. DOI: 10.1186/s13643-021-01745-y PMID: 34174957

Yin, R. K. (2017). *Case Study Research and Applications: Design and Methods*. SAGE Publications.

Zamora, J., Abraira, V., Muriel, A., Khan, K., & Coomarasamy, A. (2006). Meta-DiSc: A software for meta-analysis of test accuracy data. *BMC Medical Research Methodology*, *6*(1), 31. DOI: 10.1186/1471-2288-6-31 PMID: 16836745

Zhai, C., & Guyatt, G. (2024). Fixed-effect and random-effects models in meta-analysis. *Chinese Medical Journal*, *137*(1), 1–4. DOI: 10.1097/CM9.0000000000002814 PMID: 37612263

About the Contributors

Abdelrahman Elsayid (Abdelrahman M Makram) is an obstetrics and gynecology resident in Germany and a Master of Public Health holder from Imperial College London. He has had a great interest in medical research since his second year of medical school. He has published more than five dozen papers in multiple fields, including orthopedics, women's health, pediatrics, infectious diseases, and public health. He has also published a book chapter about the use of artificial intelligence in advancing medical biology. He serves as a peer-reviewer at more than twenty journals and an academic editor at three prestigious journals. His research mainly focuses on public health and women's health.

***

Mohammad Amin Akbarzadeh earned his MD from Tabriz University of Medical Sciences. Passionate about evidence-based medicine, he has been a member of Cochrane since 2020, contributing to evidence-based research for global medical practice. With expertise in statistical analysis, he aims to advance knowledge in evidence-based medicine, epidemiology, and hematology/oncology.

Queen Anoma, a 27-year-old graduate of Imperial College London's Master's program in Public Health, specializes in enhancing global health outcomes, with a particular emphasis on Europe and Africa. Passionate about combating health disparities and promoting disease prevention, she conducts research and leads community outreach and advocacy initiatives in the field of obesity and adolescent child health. Driven by a commitment to advancing health equity, she combines her expertise in public health with a passion for improving health outcomes on a global scale, particularly among vulnerable populations.

Trisha Bhathivi is an undergraduate student in the Judy Genshaft Honors College at the University of South Florida, pursuing a double major in Public Health and Biomedical Sciences. She is passionate about research and is driven by a desire to tackle global health challenges. Trisha's current research interests include oncology and cardiology.

Lam Chau is an undergraduate student currently pursuing a Bachelor of Arts in Biochemistry at Case Western Reserve University. With a passion for delving into systematic reviews and meta-analyses methodologies, Lam is deeply engaged in understanding the intricacies of organizing and analyzing vast amounts of scientific data. Known for their meticulous approach to research and keen analytical skills, Lam is committed to contributing to the advancement of knowledge in the field of biochemistry. Outside of academics, Lam enjoys exploring new scientific literature and engaging in discussions about the latest developments in the field. With a strong foundation in biochemistry and a fascination for research methodologies, Lam is poised to make significant contributions to the scientific community.

Federica Cucè received her MD from the University of Padova in 2021. She's currently working as a resident physician at the University of Verona. From 2019 she collaborates in systematic review and meta-analysis projects on diverse topics in the field of medical sciences, that have been published in international journals and presented in international medical congresses and has participated as teacher and tutor in courses on how to conduct research projects.

Mohamed Dawood is a doctoral student at Miami University (Ohio), where his research focuses on the intersection of pancreatic cancer cell motility and the immune landscape. Previously, he earned an MSc in Cancer Immunology and Biotechnology from the University of Nottingham's School of Medicine and Health Sciences. During his master's studies, he examined the role of myeloid-derived suppressor cells (MDSCs) in disease progression and patient survival within glioblastoma. Mohamed has long been interested in the complexities of the human immune system, particularly in health and disease. Mohamed has chosen to specialize in Cancer Immunology. He aspires to establish himself as a translational scientist, dedicated to developing novel cancer immunotherapies that bridge the gap between laboratory discovery and clinical application.

Toka Hassan, MD, is a dedicated physician committed to delivering patient-centered care and advancing medical knowledge through clinical practice and research.

Mohammad-Salar Hosseini, MD, MPH, Fellowship in Palliative Care, is a physician-researcher, graduated from Tabriz University of Medical Sciences. As an evidence-based medicine enthusiast and advocate, and a member of Cochrane, Iranian EBM Center–a JBI Center of Excellence, and Research Center for Evidence-Based Medicine (Tabriz, Iran), he looks forward to bridging the gap between research findings and clinical practice throughout his career in order to obtain the highest standards of patient care.

Nguyen Quan earned his MD from Hue University in 2021. He is driven by a passion to improve the health of his community. During his time in medical school, he has worked with the NGO Community Medicine (Y Hoc Cong Dong) to address the knowledge gap and provide primary care to underserved populations in Vietnam. Amidst the COVID-19 crisis, his team successfully secured a $130,000 grant "Healing Saigon," providing essential medical supplies like masks, PPEs, HFNCs, etc. to hospitals, directly aiding in patient care. He has been a working group leader at ORC (Nagasaki University, Japan) with interests in systematic reviews/meta-analysis, infectious diseases, and medical education.

Celine Tabche is a teaching fellow and postgraduate researcher in the School of Public Health at Imperial College London. She has a background in Biology from the American University of Beirut and MS in Eating Disorders and Clinical Nutrition from the University College London

(UCL). She is certified as an Associate Nutritionist and a fellow of higher education in the UK. She is also a Member of the Royal Society for Public Health and an Associate Faculty of Public Health member. Currently, working with the WHO Collaborating Centre at Imperial College assisting setting up a public health academy in KSA and in developing a competency framework for the emergency workforce with WHO. She has taught many short courses at the centre and helps the faculty in marking student assessments (MPH, GMPH, MSc). She has been working on multiple reports and articles involving emergency preparedness, rapid response team training, global burden of disease, bariatric surgery and obesity, infectious diseases, the public health system, health policies, health system resilience and development, medical education, and many more. Celine has given multiple conference talks on preventative medicine and nutrition and has over three years of experience in clinical nutritional consultation. She is an active member and Board Secretariat for the Arab Public Health Association and an international member of NNedPro. She also works with the International Association of Medical Colleges.

Sai Kaushik Upputuru is an undergraduate student in the Judy Genshaft Honors College at the University of South Florida, pursuing Biomedical Sciences and Public Health. His current research interests include genetic modifitcation of bacteriophages, ophthalmic drug delivery, and cancer research. He aims to identify research gaps in the medical field and is committed to improving human health and well-being.

Index

N

O

P

Q

R

S

T

U

www.ingramcontent.com/pod-product-compliance
Lightning Source LLC
LaVergne TN
LVHW080857300926
849162LV00003B/454
* 9 7 9 8 3 6 9 3 3 0 0 7 4 *